Go See the Eclipse

Go See the Eclipse

and Take a Kid with You

**Preparing for the
August 21, 2017
Total Solar Eclipse**

Chap Percival

Published by Bee Ridge Press
Sarasota, Florida USA 34239

Go See the Eclipse: And Take a Kid with You
ISBN: 0986197521
ISBN: 978-0-9861975-1-2 (ebook)

Cover concept by Bonnie Percival

ISBN 13: 9780986197529

Printed in the United States of America.

NOTE Every effort has been made to ensure that all information in this book is accurate and
correct. The author and publisher assume no responsibility for any injury, loss, or damage caused
or sustained as a consequence of the use and application of this book. An adult should supervise
young readers who undertake the activities described in this book. Enjoy the eclipse responsibly.
This book contains links to external third party websites. These links are being provided as a
convenience and for informational purposes only; they do not constitute an endorsement or an approval by the author and publisher of any of the products, services or opinions of the corporation
or organization or individual. The author and publisher bear no responsibility for the accuracy,
legality or content of the external site or for that of subsequent links. Contact the external site for
answers to questions regarding its content.

This book is dedicated to my family—Bonnie, Tina, Brandon, and Nadia— and the thousands of students I have had the pleasure of teaching during my career in the classroom.

Acknowledgments

Many people assisted me in the writing and preparation of this book. Much thanks to my wife, Bonnie, for reading this manuscript almost as many times as I did and for giving me countless helpful suggestions. Thanks also to Steve Dacey, Pete Abbott, Glynna Gower, Stephanie van Pelt, and Jane Porter for reading this book and giving terrific feedback. He may not be aware of it, but I want to thank Captain Johnny Walker for giving me the idea for the subtitle.

Table of Contents

Foreword

When I was growing up in Virginia, my mother spent her free time reading science fiction. Today we would call these science-fiction books by noted authors (such as Robert A. Heinlein, Isaac Asimov, and Larry Niven) "classics." The artwork in these books was exciting—spaceships, astronauts, planets, and aliens. I simply could not resist picking them up and reading them.

I loved the idea of traveling through space to exotic and intriguing places with advanced technology that solved problems, like how to go faster than light. Many of these books were a unique blend of fiction, space opera, physics, space sciences, sociology, and other disciplines and told of a better future for humankind, describing civilizations largely free of many of the social issues that affect us today.

In addition to being an avid reader, I was very active in the Boy Scouts and found myself frequently outdoors and away from the city lights. During these camping trips, I often looked up at the stars and constellations with wonder. I could generally find the North Star (Polaris) in the heavens but simply could not identify all of the constellations in stargazing guides of the late 1970s.

As I went through my high-school years at Norfolk Christian High School, I maintained my strong interest in science and science fiction, which culminated in taking physics my senior year. My instructor was a self-proclaimed "science nerd" who was bound and determined to take a small group of sometimes enthusiastic students and teach us the joy of physics. During my first semester, I discovered that my instructor, Chap Percival, was an astronomy enthusiast. I also found that he was willing and eager to teach me astronomy for credit as independent study.

I used to spend multiple evenings looking at the stars from the roof of our high-school gym—now with a knowledgeable person who knew the names of the primary stars and constellations. Chap also introduced me to observational astronomy with an eight-inch Schmidt-Cassegrain telescope. Some of my fondest memories from my senior year in high school are of being outside with the telescope and seeing the moons of Jupiter, the rings of Saturn, the small disk of Uranus, and deep-sky objects. In short, the stars and planets I had enjoyed reading about in science fiction were now real—and I could find them in the night sky!

I went off to college and studied chemical engineering and then got married and had children. As I was wandering through a bookstore several years ago, I came across an astronomy magazine and decided that it was time to get back into observational astronomy. I bought a telescope and have had the opportunity to share the joy of seeing Saturn's rings for the first time with one of my own children.

We have a unique opportunity in "backyard" astronomy coming up in 2017, and Chap Percival has written a book that anyone can use to help experience one of the truly magnificent astronomical events of our generation, a total eclipse of the sun. I urge you to read through the book, grab a child, and then go experience a total eclipse with family and friends. Afterward, when the astronomy bug has bitten you, buy some backyard astronomy books, grab some binoculars, and enjoy the beauty and majesty of the skies.

Onward and upward to the stars!

Kirk Schulz, President
Kansas State University

Introduction

Life is a journey, not a destination.

I would like to invite you to join me on a journey. This is no make-believe or imaginary journey. It's an actual pack-your-bags-and-head-out-the-door kind of journey. Many of you will be able to drive this journey. Some will choose to fly, take a train, or possibly a ship. A very few of you will be able to just step outside your front door. This journey has a destination, a purpose, and, best of all, will provide most of you with a once-in-a-lifetime experience. However you do it, please, please, please join me on this journey.

Sound intriguing? Pique your interest just a little bit? Well, good. You see, I have been on similar journeys before. Five times. Each one was memorable in its own way. A portion of this book is devoted to these five travels. I want you to see how valuable and worthwhile these trips have been and instill in you a desire to take your own journey. In addition, I am sure that most of you have a school-age child in your own family or sphere of influence. My goal is to not only fill you with the desire to see this eclipse but to take this school-age child with you.

As you may have figured from the book's title, my journeys involved going to see an eclipse, specifically, a total solar eclipse (TSE). (NOTE: I use those initials throughout the book.) While a TSE was the catalyst for my five journeys, the eclipse became part of a much larger package, with the end result being a huge hook on which to hang memories. Some examples of universal memory hooks (for Americans) are 9/11, when the World Trade Center was attacked; the day JFK was shot; and, for an earlier generation, the attack on Pearl Harbor. Everyone alive when those events happened know exactly where they were.

In addition to those, we each have personal memory hooks through events like weddings, births, and other similar occasions. By taking this journey, you will give yourself another such hook.

Solar eclipses occur every eighteen months, on average, somewhere on the earth. Why is this eclipse so special? Why am I writing a book in preparation for this particular eclipse and not any of the other five I have journeyed to see? Great questions. And here's the answer. You see, the continental United States has been missing out on this greatest of celestial spectacles for decades. As a result of the peculiar rhythms of the orbits of the earth around the sun, the moon around the earth, and the rotation of the earth, we in America have not seen this type of eclipse on US soil (the forty-eight contiguous states) since the 1970s. It has occurred in lots of other places, just not here.

If you are under the age of forty and have not traveled outside the United States, there is no way that you have experienced one of these TSEs. That is more than half the population of the country. Most people over forty have not seen one either. I estimate that 90 to 95 percent of the population of the United States has not seen a TSE. That is upward of 280 million people, and they have no idea what they have been missing.

Some of you may protest and say that you have, in fact, seen a solar eclipse. You remember as a kid having the shoe-box viewer or small telescope and projecting the image of the sun, with a bite taken out of it, onto the screen. While that is almost certainly true, what you witnessed was a partial solar eclipse, not a *total* solar eclipse. They are not the same. Not nearly. Trust me on that.

Your next questions may be, "When *is* this eclipse? How much time do I have to prepare? What do I need for it? How do I get ready? Where do I go?" More great questions. Well, you have some time. As I write these words, you have about two and a half years to prepare for your journey.

The good news is that the eclipse drought that Americans have been experiencing is about to end. On Monday, August 21, 2017, a TSE will be visible in the United States. A TSE that travels west to east, coast to coast, Pacific to Atlantic across the entire country is rare. It last occurred in 1918. This will be a big deal. Astronomers, professional and amateur; eclipse chasers; adventure seekers; and the curious from around the world will be coming to America, just to witness this event. TSEs are spectacular events—no hyperbole here. This, combined with their rarity, tells me that a bit (OK, maybe more than a bit) of preparation is in order. I want you to go see this eclipse. And I want to help you get the most out of the experience. This will be a memory-making event for you, your family, and friends. I want to help you prepare for it.

This book is aimed at nonastronomers. It will appeal to anybody who wants to experience something new and different. It is for people who might not know much about astronomy but enjoy looking up at the sky. It is also for those who want to learn more about the heavens and are amazed by the universe we live in. Because of this, I will not be as rigorous and technical in my definitions and explanations as I would be if I were writing for a more advanced audience. This will, hopefully, improve readability. For the astronomically savvy among you, please understand and excuse the lack of rigor.

I want to not only prepare adults to see this event but also school-age children—hence the subtitle. Because of that, I have included many activities and what could be called "teachable moments," in which an adult using simple props can help a child understand some of the concepts of space and motion related to eclipses.

Throughout this book, I will regularly remind you of two things:

1. **Never look at the sun without proper eye protection.**
2. **Go see the eclipse, and take a kid with you.**

PART ONE
ECLIPSES

1

What is an Eclipse?

Before our journey can begin, I have to answer the question, "What is a total solar eclipse, and why is it worth the time and effort to go see one?" Since this section can be a bit technical, I have decided to handle it two ways. The first uses the KISS (Keep It Short and Simple) method. The second will contain more detail and complexity and provide more information about eclipses without being too technical.

Have you ever played or seen a toddler play peekaboo? If so, you could say you have seen an eclipse. The hands eclipse (or hide) the face. Another place you may have seen or experienced this is when driving toward the rising or setting sun. Without clouds to hide the sun, you must resort to using the visor in the car or your own hand held up to block (or hide) the sun.

In the most general terms, an eclipse in the sky is an event where one object passes in front of another, more distant object and hides it for a period of time. The duration of that hiding may be as short as a few seconds, up to an hour, or possibly more, depending on the objects involved.

From the earth, we can see two kinds of eclipses: lunar and solar. A lunar eclipse can occur only when the moon is full, and a solar eclipse can occur only when the moon is at the new phase. The order of the three objects in a lunar eclipse is sun, earth, and moon. During a partial or total lunar eclipse, the moon enters the shadow of the earth and the normally bright full moon becomes dark. The earth's shadow is wide—wider across than the moon—so exactly how dark the moon becomes depends mostly on how far into the earth's shadow it travels.

A solar eclipse happens when the moon is new and passes exactly between the sun and the earth. A solar eclipse does not happen every month even though there is a new moon every month because the moon is usually either a bit above or below the earth.

On a side note, imagine if we had a spaceship that we could fly a few hundred thousand miles. That would mean that we could go up any time we wanted and enter the shadow of the moon and see a solar eclipse. That may happen someday, but for now we have to wait for the shadow to touch the earth.

The moon is smaller than the earth, and thus, its shadow is much smaller than the earth. It is, in fact, only fifty to one hundred miles wide where it darkens the earth. The earth is eight thousand miles in diameter. That means only a tiny portion of the earth actually experiences the total eclipse. The partial shadow (penumbra) is much bigger, so more people have seen a partial solar eclipse. While they are interesting, they are not spectacular. Only total solar eclipses can proclaim that.

Eclipses: More Detail

One example of an eclipse is when the moon passes in front of a star or planet. (Astronomers call these events "occultations," which is another word for hiding.) These hidings typically last for a couple of hours.

What causes this eclipsing? Two things: first, the moon is the nearest celestial object to the earth and, second, the moon moves in an orbit around the earth, about once a month. There are faint stars that the moon hides every day, but to hide a bright star is much less common, and to hide a planet is rarer still. Technically a planet can pass between the earth and a star (not very common) or even a more distant planet (very rare).

Here is a quick and dirty way to make an eclipse with two items you likely have, a quarter and a dime. Take the quarter in one hand, and hold it between your thumb and first finger so you can see its circle shape (not edge on). Now take the dime in your other hand, holding it between that thumb and first finger, and position it about halfway between the quarter and your eye. Go ahead. Try it. Close one eye, and look at the quarter. Gradually move the dime between your eye and the quarter. If the dime is at the right distance, you will see the quarter disappear. You have just made the dime eclipse the quarter.

The earth experiences two kinds of eclipses: solar and lunar. The sun is bright, and any object in the sun's light casts a shadow. We have all felt the relief of entering the shade (or shadow) of a tree on a hot summer's day. The moon orbits (or revolves around) the earth. Since the earth and the moon cast shadows produced by the sun, it is possible for the earth to pass through the shadow of the moon and the moon to pass through the shadow of the earth. When the moon enters the earth's shadow, we say the moon is eclipsed, or hidden (a lunar eclipse), and when the earth enters the moon's shadow, it hides, or eclipses, the sun (a solar eclipse).

The earth's diameter is about four times that of the moon's, resulting in the earth casting a bigger and longer shadow. When the moon enters the shadow of the earth, it can take an hour or more to pass all the way through and come back out into the sun's light. This kind of eclipse can only occur when the moon is full. The shadow of the earth is dark but not black. The earth's atmosphere scatters and refracts the sun's light that enters it, and some of this light enters the shadow, brightening it a bit. The result is that a fully eclipsed moon is usually a shade of red or orange.

These events can last for an hour or more and are cool to see. Everyone should see at least one of these. But they happen at night, sometimes in the wee hours and are, therefore, difficult for small children to see.

Earlier I said that the moon's shadow is smaller than the earth. How much smaller is it? Small enough that when the moon passes between the sun and the earth, the diameter of the shadow the moon casts on the earth is only about 1 or 2 percent of the earth's diameter. As the moon moves in its orbit, the shadow moves in a line, darkening a small area. The total area of the earth that is in the shadow of the moon and will experience this TSE is about two-hundredths of 1 percent of the earth's area. Figure 4 (at the end of this chapter) shows the path of the August 21, 2017, eclipse's umbra (full shadow) as it moves along the earth's surface. It crosses oceans and land and, for most of its length, is sixty to seventy miles wide. Other eclipses will vary in size, but this eclipse is near the average.

TECHNICAL NOTE: A word about diagrams of the earth, sun, and moon. You will rarely, if ever, see a true scale diagram of these three objects. This is because the three objects are very different in size and extremely far apart. Compared to how far apart they are, they are miniscule. The usual diagram of the earth and the moon may show the correct size relation, but not the distance. Any scale diagram of both size and distance would have a lot of white space, something most publishers do not like to see. I don't have that constraint, so here is my explanation of just how big the earth-moon-sun system is.

Figure 1. *The moon and earth in scale, both in diameter and distance apart. The moon is about thirty earth diameters away from the earth.*

In figure 1, the moon's diameter is about one-fourth that of the earth's. It is pretty clear from this diagram that there is a lot of space between them. Anybody with an e-book can zoom in on this page, but the diagram was made to have the moon and the earth about six inches apart. If your image is not at that scale, try to adjust it. If you can't do that, just imagine that they are. Now the scale is set. How big is the sun on this scale, and how far away would it be? Take a beach ball that's about twenty inches in diameter. That is the size of the sun on this scale. Now take that beach ball, and walk to the left, about two-thirds the length of a football field (two hundred feet). *That* is where the sun would be, using the scale of the earth and moon on this page. There is a reason why it's called outer *space*.

There is a fancy word for the arrangement that produces an eclipse: syzygy (SIZZ-uh-gee). (I have always wanted to use that word in a sentence.) It means the alignment of three celestial bodies in space along a straight line. It can certainly refer to the sun, a planet, and one of the planet's moons. In the case of a total solar eclipse, it consists of the moon lining up between the sun and the earth. This can only occur during the new moon phase. Some of you may wonder, since there is a new moon every month (approximately), why is there not a solar eclipse every month? The answer requires some three-dimensional thinking, not something we do very often. Please bear with me as I explain.

The earth orbits the sun in a plane called the ecliptic. The moon orbits the earth in a plane. But those two planes are not the same. They are close to each other but are tilted with respect to each other by a mere five degrees. That is all it takes. The two planes do intersect, but that means the planes are together only along a single line. If you lay an open book on a table, lay the pages flat (forming a plane), and pick one of the pages up so that it tilts a little, you have two planes that intersect along a line, the binding of the book.

An eclipse can happen only when the moon, as it orbits around the earth, crosses the plane of the earth's orbit around the sun at the same time as the line of intersection of their orbital planes is aimed at the sun. (See what I mean about three-dimensional thinking.) This may be hard for you to follow. Figure 2 portrays the arrangement pretty well. Fortunately it is not essential for your understanding about eclipses. The key fact here is that eclipses can't occur every month because of this arrangement. The moon is usually either above or below the earth. Please note that this diagram is approximately to scale. The shadow of the moon is very skinny. Most textbook diagrams of the moon-earth system with shadows do not look like this. This is the real deal.

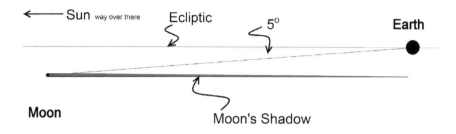

Figure 2. *The moon's shadow misses the earth most of the time.*

Here is another way to think about the scale of the moon's shadow. Take two, ten-foot-long pieces of string, and lay them stretched out side by side. Seriously, do this, especially if you have school-age children. Have one person stand at one end of the two strings and pick them up and hold the ends together. Another person should grab the other ends, pick them up, and stretch them pretty taut, while the first person holds steady. The second person should separate the two ends and hold them about one inch apart. The strings will then make a very skinny angle. That angle is a good representation of the cross section of the moon's shadow—long but skinny. Now imagine that a total solar eclipse occurs only when the skinny end of the angle actually touches the earth in space. On this scale, the earth is about four inches in diameter and presents a really small target for the moon's shadow as the moon revolves around the earth. It is amazing that we have any solar eclipses at all.

NOTE: For more activities about eclipses, see Part Four.

More people have seen a lunar eclipse than a solar eclipse. (Refer to figure 3 to see what one looks like.) The reasons are as follows. The earth is bigger than the moon and has a bigger shadow—big enough that the entire moon fits in it. When the moon enters the shadow of the earth, every person on the half of the earth facing the moon can see the eclipse in progress. Each person will see the moon in a different position in the sky, but they will all see the moon entering the shadow. In addition, since a total lunar eclipse can last for over an hour and the earth rotates fifteen degrees per hour, a portion of the earth's population will not see the start of the eclipse but will see the last part of the eclipse after the moon rises at their location. This is not important in and of itself, but it means that, on average, slightly more than half the world's population can see a given lunar eclipse. Contrast that with the earth entering the much smaller shadow of the moon for a solar eclipse. As I mentioned before, the entire shadow is much too small for the earth to fit in. As a consequence, only those people in a very narrow path get to see the entire sun blocked out by the moon. But those lucky people get some show.

The other reason fewer people see a solar eclipse is related to the fact that over 70 percent of the earth's surface is covered with water. That means, on average, 70 percent of total solar eclipses are visible only from the water, where few people live.

Figure 3. The April 15, 2014 total lunar eclipse. Notice that the earth's shadow is not uniform in color or darkness. The center is near the top of the moon, where it appears darkest.

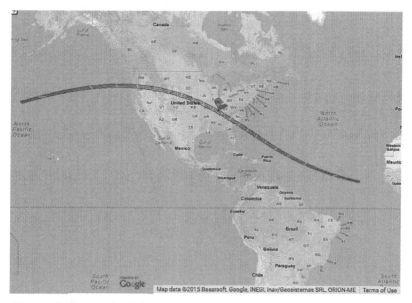

Figure 4. The path of the August 21, 2017 total solar eclipse. Only people in those locations between the blue lines can see the total eclipse.

2

What does an Eclipse Look Like?

Well, you may ask, what does a total solar eclipse actually look like? Perhaps a better way to ask it is: what is the eclipse-viewing experience like? And believe me, it is an experience. Let me attempt to describe the day of a total solar eclipse.

Depending on where you are in the path, this eclipse can start as early as 9:00 a.m. in coastal Oregon or as late as 1:10 p.m. on the South Carolina coast.

You should keep track of the time so you know when the partial eclipse begins. You can't see this happen with the naked eye, even with proper viewing glasses. (More about eclipse viewing safety later on.) If you have a telescope properly outfitted with a solar-viewing filter or a projection screen, what will you see? The round sun will have a tiny nick taken out of it as the moon first touches it. The size of the nick will increase over the next hour, without having much effect on the way the sky looks. Once the sun is more than half covered, the sky begins to take on a slightly altered appearance both in color and brightness. It looks a bit like dusk, but in a slightly disconcerting way, since it is, after all, the middle of the day.

The sun is now clearly a crescent, ever gradually shrinking. It seems to take forever for the sun to be completely covered by the moon, but the reality is that it will take only about an hour and a half. In the last half hour of the partial phase, the sky gets darker and the surroundings and the landscape become harder to see. If you look around, you may notice that the horizon looks like the color of dusk in every direction. Now that never happens without an eclipse.

In the final few minutes before totality, time seems to speed up. What was a nice leisurely pace becomes more up-tempo. More things are happening at the same time, making it difficult to notice it all. Don't worry about this, especially if this is your first eclipse. You want to focus on the main show, the eclipse itself.

As the last piece of the crescent-shaped sun shrinks and fades away, the incredible flash of a diamond ring will appear for several seconds before succumbing to the relentless march of the moon across the sun, hiding the rest of the photosphere. And then it quickly gets much darker.

The eclipse appears as if there is a black hole in the sky surrounded by the pale white flames of the solar corona. The streamers (white, flame-like projections from the surface of the sun) of the corona are arrayed around the circumference of the moon (and sun behind it) in a pattern unique to every eclipse. They don't move during the eclipse, but they can appear to shimmer. You may also see some small, reddish pink areas called prominences around the moon (and sun). These are eruptions on the sun's surface that happen all the time but are usually hidden from our view by the dazzling photosphere. Many are larger than the earth.

The few minutes that the eclipse lasts pass much too quickly, and before you know it, the moon begins its retreat from the sun and a second diamond ring appears—this time on the opposite side of the sun from where it first appeared. Then begins the mirror process: the skinny crescent of the sun grows, the surroundings brighten, and the air warms back up to its original condition. It is truly anticlimactic, almost disappointing, after such a high. But what a high. And this part of the eclipse does give you a chance to reflect on what just happened. Immediately a thought comes to mind (my mind anyway): "I want to see another. When is the next one?" As each eclipse day is different—weather, location, time of day, crowd, etc.—so is each viewing experience. I will share more of my personal observations and feelings in Part Two.

PART TWO
ECLIPSE MEMORIES

3

Childhood through College

I have been involved in astronomy for many years, mostly as an educator in the classroom and in the planetarium. Although my interest and passion for astronomy blossomed in adulthood, they were planted during childhood, and then watered and nurtured throughout my growing-up years. I owe a great debt to my father for planting the seed and the early watering. He passed away when I was fourteen. He never saw the fruit of his efforts.

The youngest of four boys, I grew up in a home with a stay-at-home mom and a father who loved learning. In times of quiet, amid the mayhem inherent with four boys in one house, I remember watching the bird feeder we had hanging in the willow tree outside a dining-room window. I observed cardinals, chickadees, blue jays, finches, and woodpeckers. We lived next to a wooded lot, where I climbed trees and explored as much as I was allowed. One morning my mother called me to come see a doe standing next to our swing at the edge of the woods. I enjoyed nature from my earliest recollections.

My father was a pastor, and our home was full of books—all kinds of books. A few were related to science, but science was not my father's first love. History, languages, and biblical topics were where his heart was. I recall one title, *Six Thousand Years of Bread* by H. E. Jacobs, which I read as a young boy and still have today. Through my father's influence, I grew to love reading. But even before that, there was a wonderful memory-making event.

It is the earliest eclipse-related and, in fact, astronomy-related event that I can recall. It occurred in what I now know was 1954, when I was six years old. In fact, I can date it exactly to June 30, 1954. I can do that because, even though the frequency of solar eclipses at a given location is rare, the timing of eclipses is precisely known. We lived in Clark's Summit, Pennsylvania, and there was only one solar eclipse visible from that town in the few years between my birth and our move away from there. This is an example of a memory hook, an event so profound to the individual that it produces a memory that endures for life.

Here are some details I remember. My father was an amateur photographer. Now, that designation, *amateur*, may not convey the level of his involvement in photography, unless you know that it derives from the Latin *amare*, meaning "to love." There was one room in our house that served as his office/study/workroom. My memory of the room is not well defined, though, because I was not allowed in there without adult supervision. In any event, my father had five or six cameras. He developed and printed his own photos, meaning there was an array of chemicals, jars, containers, racks, trays, enlarger, and other associated gear.

That morning (June 30, 1954), he got out some fully exposed and developed (that is, very black and very dense) black-and-white negatives. He held them up to the sun for me to look through, and I remember seeing a white crescent shape where the sun was. My memories are dim (remember, I was only six), and while I don't recall a moment of life-changing inspiration, it was, obviously, impressive enough that I remember it all these years later. My dad succeeded in making a big impression on me.

The seed had been planted.

A few months after that, we moved to New York City. Now, the city was not (and still is not) a hotbed of astronomical viewing—far from it. The city lights make it very difficult to do any serious sky watching. Nevertheless, I remember my father occasionally taking me to the roof of our apartment building and pointing out the few stars and groups that were visible (the Big Dipper, for one). He also had a spotting scope that we used to look at the moon. Very cool. I enjoyed those times with my dad.

The seed was being watered.

Then there was a total lunar eclipse on March 13, 1960. I was not aware that an eclipse was occurring that evening, but I remember being startled (and pleased) by the orange moon suddenly visible between the tall buildings when I positioned myself on an east-west street. As I walked along the sidewalk, I could keep the moon in view. Over the next hour or so, I would look for the moon to see transitions as the earth's shadow slowly swept across it.

The other strong influence in my astronomical background during this time was the Hayden Planetarium at the American Museum of Natural History. I truly loved this place. I went as often as I could, which, while not weekly or even monthly, was enough to keep my interest in the heavens alive. I will admit that I cannot remember one show or even a detail from one. There wasn't a show that I did not like. After the programs, I would linger and just stand mesmerized by the wonder of it all. This was inspirational stuff. I simply cannot say enough good things about the influence that the planetarium and museum had on me. This bore special fruit later in my life. (I want to thank all those writers and presenters and acknowledge the influence they had on my life. Thank you.)

The plot where the seed had been planted was being watered, weeded, and tended to.

Another nurturing event happened in July 1963. Then living in Virginia, after my father had passed away, I recall being excited to hear about a partial solar eclipse visible from our town. It was total in Canada and New England, but a partial eclipse was better than none. Since the sky was clear that day, I was able to note the progress of the sun as the eclipse advanced but did not have the equipment to do any serious observing.

In 1964 I entered college. Because of my mathematics and physics double major, my time was spent in other pursuits, and my astronomical inclination was temporarily shelved.

The watering was on hold, and the seed went dormant.

4

Virginia

After graduating in May 1968, I began what became my teaching career in rural Ohio. This was an amazing time for a physical-science teacher to be alive. During my first year of teaching, the Apollo 8, 9, and 10 missions went to the moon, albeit without landing. (After the Apollo 1 tragedy of January 1967, with the deaths of Gus Grissom, Roger Chaffee, and Ed White, the Apollo missions had been put on hold until the issue of an emergency escape was addressed.) Those three missions, most especially Apollo 8, brought back to life the hope that we Americans would be able to land men on the moon and safely return them by the end of the 1960s, as President John F. Kennedy had announced in 1962. These events reawakened my interest in the sky. In 1969, I remember impressing my future wife with my knowledge of the night skies, taking her outside and pointing out several stars, constellations, and asterisms. This would have been difficult from a city, with its bright lights making the stars hard to see. We were in a rural location where the skies, on clear nights, were dark and filled with stars.

During the 1969–1970 school year, I learned of the March 7, 1970, total solar eclipse that would pass through my hometown in Virginia. My first thought was to take time off to drive down to see the eclipse. I had started an astronomy club and so decided to take a couple of students with me. I petitioned the principal to designate it a field trip, and he said he would have to present it to the school board. When he did, the board turned it down, apparently thinking I just wanted a paid holiday. I asked to speak to the members of the board in person, sharing with them the significance of the event and the impact it would have on the two students. They changed their minds and gave their blessing for the trip. In early March, I drove to Chesapeake, Virginia, with Bonnie, my bride of four months, and two high-school junior boys from the astronomy club.

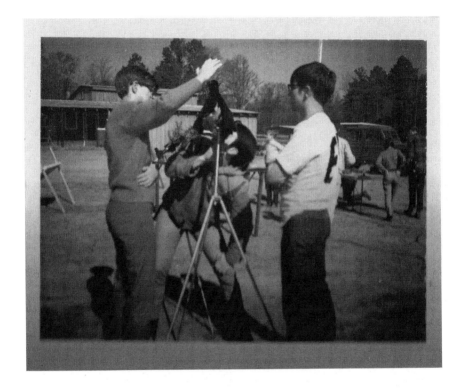

Figure 5. *The author and his students looking through binoculars mounted on a tripod at the TSE of March 7, 1970, in Chesapeake, Virginia. (Photo by Bonnie Percival.)*

Since this eclipse was in my hometown of Norfolk, Virginia, I knew exactly where I wanted to set up to view it: the Triple-R Ranch in Chesapeake, Virginia. This was the last time I would be that fortunate—every trip after was to a place I had never been before and knew little about.

The day of the eclipse, the sky was clear and blue. We had a great time. Students from other schools were there, and the excitement was palpable. I was so glad that I had the chance to see this great spectacle and bring two students to see it. But I learned how poorly prepared I was to view, and especially photograph, the eclipse. The only camera I owned at the time was a Kodak Instamatic, not the best camera to photograph an eclipse. I had borrowed an 8mm (millimeter) film camera to bring, but after arrival, I learned from more knowledgeable people that the image of the eclipse on the film would be too small to be easily seen. I jury-rigged it to a pair of binoculars I had brought along to magnify the image. (It seemed to work, but when I had the film developed once we got home, the image was out of focus.) I was nearly frantic, trying to get something tangible from the event. I was so worried about getting documentation of the eclipse to show the school board back home that I almost missed seeing the eclipse itself. But see it I did—grand, glorious, and majestic in the dark-blue midday sky. Cheers arose from everyone there, as if in tribute to the disappearing sun. Exhilarating, enthralling, and beguiling all at the same time. All in all, it was a wonderful experience. I was definitely psyched.

It is safe to say that this event fully awakened in me what had been a dormant interest in total solar eclipses.

The seed planted long ago finally germinated and took root.

5

Canada

I wanted to see more total solar eclipses. In fact, I couldn't wait to see another one. In researching TSEs, since there were no personal computers or Internet back then, I used the library a lot. I saw that there was a TSE scheduled to occur in Canada on July 10, 1972. It provided the next likely (meaning not too far away and on dry land) opportunity to see one. I simply did not have the resources to travel overseas to see one at that point.

In the early seventies, the way we got detailed information about a particular eclipse was through a document called a circular and published by the United States Naval Observatory. After requesting and receiving my copy for this eclipse, I pored over the maps and data, looking for the location with the best chance of seeing the eclipse and with the longest duration of totality. The two best places, logistically and for viewing purposes, were the Gaspè Peninsula of Quebec and Nova Scotia. I selected the Gaspè (the duration of the eclipse was a few seconds longer there) and wrote to the postmaster of Cap-Chat (Cape Cat), a small town on the center line of the eclipse. The postmaster forwarded my letter to a person who was heading a committee that was set up to assist viewers looking for a place to see the eclipse. Since I was a teacher and was bringing students, they let us use the grounds of the school in the heart of town to pitch our tents. They also let us use the restrooms and other facilities of the school. Quite a coup.

I had never been to Quebec, nor did I know anyone from there. All that I knew about it had come from reading. When we arrived and set up camp, we soon discovered that only two people in town were fluent in English (the people in charge of the committee). In shops we quickly learned the international language of pointing to the desired item and writing the prices down. And smiling—always smiling.

Cap-Chat was a delightful town on the Saint Lawrence River. At that location, the river was wide enough that you could not see the other side, making it look like the ocean. Just outside the town and overlooking the river was a rock formation that bore a resemblance to a cat sitting on its haunches, hence the name of the town.

I was determined to make a better effort at viewing and photographing this eclipse. I had a refracting telescope with an adapter for my camera so that I could take pictures through the telescope. I also had a four-by-five-inch Speed Graphic camera that had been my father's. I gleefully anticipated getting some pretty good photographs. For the next three days, we prepared for the eclipse, anxiously watching the weather. We tested our telescopes, cameras, and other equipment for full function. The days were beautiful, clear, and pleasant. At night we set up the telescopes and invited local children and families to come look through them. One person in our troop spoke French, which helped with communication. We definitely became the talk of the town. In the morning we would wake to the aroma of freshly baked goods from the nearby bakery. A memorable experience, no matter what happened with the eclipse.

The day of the eclipse dawned crystal clear, but by the time the partial phase began, clouds were building. For the next two hours, the sun played peekaboo through holes in the increasingly thick clouds. One minute before totality, the clouds blocked the sun completely. We were shut out of the eclipse. What a letdown. We did the best we could. The ambient light dropped dramatically. Birds quieted and went to roost, and an eerie color surrounded us. It was over far too quickly. I found out later that if we had gone to Nova Scotia we would have seen the eclipse. Oh well. We left the next day, disappointed but with no regrets. The trip was memorable and worth the time and effort. This would not be the last eclipse journey I would take.

However, it would be a long time until the next one. Between there being no nearby eclipses that we could drive to, raising a daughter, and being a teacher, it would be twenty-six years until I would travel to another one. I had thought about going to see the eclipse in February 1979. It was in the Pacific Northwest, so technically I could drive to it, albeit 2,500 miles. But the reality was that the weather conditions were not favorable, and it would be cold. Plus, I was a teacher and hated to miss school. It is such a pain to make substitute lesson plans.

Remember:

1. **Never look at the sun without proper eye protection.**
2. **Go see the eclipse, and take a kid with you.**

Figure 6. Our eclipse viewing equipment on the school grounds in Cap-Chat, Quebec. Photo by Bill Willaford.

Figure 7. This is a map of the Gaspè Peninsula that I obtained from the Quebec department of Tourism. I drew the boundary lines and the center line of the path of totality of the eclipse of July 10, 1972.

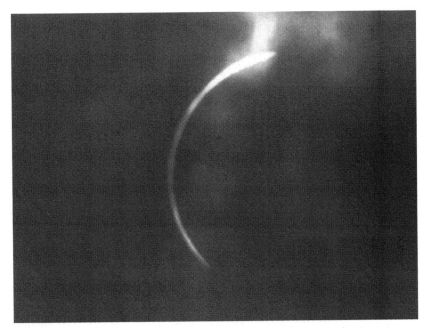

Figure 8. This is the closest we got to seeing the TSE of July 1972. Missed totality by two minutes.

6

Aruba

By 1998 I was teaching astronomy at the high-school level in Sarasota County, Florida, and my daughter was a high-school senior, soon to enter the workforce. My wife and I took the plunge and, with another couple, booked a Caribbean cruise with an itinerary designed to view the TSE of February 26, 1998.

This was the middle of the school year. The Internet and World Wide Web were young, and I came up with the idea of posting daily commentary of my trip for my students to follow. Several of the islands had newly created businesses called Internet cafés that I could use to transfer files. The term "blog" was yet to be coined, but I had a website I could use to post daily updates. We flew from Saint Petersburg, Florida, to San Juan, Puerto Rico, and boarded the *Dawn Princess* with over two thousand of my closest friends (well, at least birds of a feather), most of who were along to see the eclipse. I was in astronomy-nerd heaven.

The ship called at St. Thomas, Dominica, Grenada, Venezuela, and Aruba. We quickly established a routine that went pretty much like this: We would spend the day on shore doing touristy things. Back on the ship, I would use a laptop to create a file for my website. At the next port, I would find the Internet café in town and upload my report. I had no digital camera, so I could only upload text. Only then could we continue with our sightseeing. The Internet cafés were usually not in highly visible locations. Some were still under construction, in cramped quarters, and sometimes hard to locate.

In St. Thomas, I had to go through alleys and up stairs to a tiny room with three stations set up for use. Most had been in business only a couple of months. But, by golly, they were on the World Wide Web.

After uploading my web post, we did some shopping and then headed on a fast ferry ride to St. John for some snorkeling at the underwater park. At Dominica I had secured the services of a guide and did some bird-watching in a preserve while my wife stayed in town and did some sightseeing. On Grenada, the Spice Island, we walked around St. George's and visited the large market, where we stocked up on some of the wonderfully fragrant spices for consumption after the trip.

The most challenging file transfer occurred in Venezuela. There were no Internet cafés that I could locate, but on the way off the ship, I found a phone station on the pier. I paid the proprietor for the use of the phone to dial up and transfer the files. (My laptop had a PCMCIA modem/Ethernet card, for those who care.) It took ten minutes to upload the minimal files I used for this transfer. Since he charged five dollars a minute for the call, the total cost for that one upload was fifty dollars—way more than I had allotted. But other than that, the use of Internet cafés worked pretty well, and people back home could see how the trip was progressing.

While in Venezuela, we went on a coach-bus tour of Caracas and its surroundings. We stopped at the Murano-style glassblowing factory and saw workers make some of their specialty glass clown figurines. On the return to the ship, we passed a steep hillside covered with *ranchitos*, little homes of the poor, seemingly stacked one on top of one another up the steep slope. A reality check, for certain.

The morning of the eclipse, we pulled into Oranjestad, Aruba. The cruise line gave passengers the option of staying on board and having a longer-duration view of the eclipse at sea or going ashore. I wanted to be on land for the stability needed to take photographs of the eclipse. As soon as the gangway was ready, our party of four disembarked from the ship, each one helping me lug a small telescope, camera, and accessories, along with personal-viewing paraphernalia and bags. As we scouted the streets for a place to set up, I was slightly disconcerted to see the ship leaving the pier to travel to the center line of the eclipse some fifteen miles off shore. Cruise-ship passengers are always given a healthy warning about missing the ship's departure at any of the ports of call. I mean, I knew the ship would return after the eclipse to retrieve those truly dedicated eclipse photographers, but still, it was a bit unnerving.

There was an initial air of uncertainty that surrounded us as thin clouds threatened to grow and block the eclipse. But as the moon covered the first piece of sun, the temperature dipped just a bit, making the clouds flee the scene and leaving us with a clear blue sky. As the midday twilight advanced, the quality of light changed. It got darker, but more than that, the colors changed. This can only be experienced during a total solar eclipse. The result is a 360-degree panoramic view of dusk, with the horizon showing a palette of yellows, oranges, and reds no matter where you look. I tore my gaze from the horizon and looked up just in time to see the last piece of bright sunlight being obscured by the moon in that fantastic transition known as the diamond ring, my wife's favorite part of the eclipse. The duration varies from eclipse to eclipse because the exact profile of the moon is uncertain. For instance, a deep valley on the edge of the moon may cause the sun to linger a few seconds more than usual. In any event, the diamond gradually fades, and the glorious corona makes its appearance.

"Corona" is Latin for "crown," and the name is quite apt. The stunning view you encounter during the total eclipse is like one from a painter with an eye for the dramatic. The sky high above the horizon is much darker. The planets and brighter stars become visible. Superimposed on the faded background is the small, dark hole in the sky surrounded by a ring of pearly white flames, some stretching three and four times the moon's diameter.

I went to the viewfinder of my Minolta SR-T 101, attached to the Celestron 5+ telescope, and snapped a bunch of pictures. The focal length of the telescope was 1,250 millimeters, which meant that the image of the sun's disk was too big for me to get the full extent of the corona in the frame. Darn. I had to be content with some killer photos of the diamond ring, the inner corona, and some prominences. (With each eclipse, you learn a little that might be useful during future eclipses—if you can remember.)

Immediately following the end of the eclipse, I rushed into a nearby (strategically so) Internet café and filed a story of my experience. Since I was using film and did not have a digital camera, there were no photos, just text. But it was immediate and felt great.

That evening the entire ship resounded in one big celebration (at least a big celebration by nerd standards). The only specific thing I recall is that as you walked along the ship's decks, everyone was smiling and cheerful, the cares of life held in abeyance for the moment. Many toasts were made that evening, and goodwill was in abundance. We had shared what would be, for some, a once-in-a-lifetime experience. But for me, this was my second eclipse (I did not count my trip to Quebec since I did not see totality), and I was ready for more.

This trip was memorable for several reasons. It was my first cruise. That alone would make it unforgettable. But combine that with the dining, shows, camaraderie of being with friends and many like-minded enthusiasts, and the ports and scenery, and everything came together to make great memories. Oh, and I got some killer photographs of the eclipse.

I was itching for more. I could not wait. Next up: Europe.

7

Hungary

Immediately upon our return home, I began developing a plan to see the August 11, 1999, total solar eclipse in central Europe. I had seventeen months to prepare and needed to make the most of the time. I decided to try to recruit other people and lead a group. I contacted several educational travel companies and looked at a bunch of catalogs. They all had many trips to England, Italy, France, and Spain, but not very many to Austria and Hungary. I finally found one that looked promising.

I contacted that company and described the trip I was contemplating. My Aruba eclipse photos convinced the owner that a dedicated eclipse trip was worth arranging. He also asked permission to use one of my diamond ring photos as the cover artwork for his company's 1999 trip catalog, which featured the eclipse trip. (My hat size increased just a bit.)

I managed to recruit enough people for the trip to fly (literally), and when a small group from Utah signed on, we had enough for our own coach. Prior to the start of the trip, at every opportunity, I reminded the company owner that we needed to be at the eclipse-viewing site early in the morning on eclipse day. There would be no reason to go on the trip if we got to that location the day after the eclipse. On these educational trips, you are usually going to experience the culture, tour some buildings, and meet some people. This can be done one day as well as the next, making my insistence on being at a certain place on a certain day outside the norm for this type of trip. The owner finally understood the need because, when we arrived at our rendezvous site in Prague, there was not one chaperone, not two chaperones, but three chaperones. In addition to the one originally assigned to our group, there was another who said that he was intrigued by the eclipse and was taking personal time to accompany us, and the third was native to Hungary (where our viewing site would be). She was a last-minute addition and was not happy about it. She enjoyed traveling to new and different places and did not want to be in her homeland when she could be somewhere more exciting. She did not understand why the eclipse was generating such a huge amount of buzz or why we were thrilled at the prospect of viewing it. In any event, it proved to be a wise move. Since Hungarian is not as widely spoken as some languages, having her along proved to be fortuitous, happy or not.

After touring Prague and Vienna, we headed to Budapest. The group toured the city for a full day while I hired a driver and car to go scout out potential viewing sites. Without actually traveling to Hungary prior to the eclipse trip, it was very difficult to determine, in more than the most general geographic terms, our location for viewing the eclipse. I had the map of the eclipse path and so knew some good areas to check (about an hour south of Budapest) but knew no exact sites. So a scouting trip was imperative before eclipse day. That morning I, one other tour member, and the abovementioned Hungarian chaperone piled into the hired car and were driven south for what I thought would be an all-day fact-finding mission. The road was good and the driver fast. I had originally thought that the Tihany area on Lake Balaton would be promising, but I heard that the government had closed the national park there for eclipse day. All those constraints told me to head for Veszprém. We found a good road that went west from there toward the center line of the eclipse. At the outskirts of Herend, we saw a building with an adjacent field. It was our first and, as it turned out, final stop in our search for a viewing site. We walked into the building and found a restaurant and hotel, Viktória by name. The restaurant was empty, as it was early in the day. We found an employee, and our Hungarian chaperone went to work. She described our situation and that we would be bringing a busload of people the next day. Would they be able to accommodate us and not take any other reservations? They agreed, but only if we would patronize the restaurant. It sounded like a plan, so we exchanged names and numbers and returned to Budapest, but earlier than we expected, I might add.

On the trip back to Budapest, the other tour member who had come along, a veteran, asked that we stop at a war memorial cemetery we had passed on the way out. We did and paused to reflect on the events that surrounded the lives of the soldiers buried there. While it was a very sobering visit, it was another memory maker in the best sense of the term.

Since our search for a viewing site had been so quick, easy, and successful, we arrived back at the hotel a few hours sooner than I had anticipated. Because the rest of the group was still on a tour of Budapest, I opted to explore the area on foot. I found an aerial tram and rode up to the top of a hill. After enjoying the view of the city awhile, I strolled around a bit and then meandered back to our hotel. The group returned, and we swapped tales of the day's adventures. Spirits were high because the next day was E-day and the weather forecast looked promising.

The total eclipse was widely publicized in Hungary and was a big deal. Postage stamps were issued commemorating it, and many took the day off. (It was August, which is a big holiday month in Europe.) I had no idea what traffic would be like or how long it would take us to get to our site. (We needed to be there by eleven the next morning.) Five-thirty in the morning came too soon, but we got up, ate, packed, and headed to the bus. We were on the road by seven o'clock. The roads were, while not empty, moving freely, and we arrived at our viewing location by nine o'clock.

That meant we had some time before the eclipse began. Many of us walked into Herend and explored on our own. Because it was a holiday, many of the shops were closed, but enough were open to make for good souvenir hunting. And the town was, well, quaint. The architecture was pleasant, and we had a delightful hour or so. As it turns out, Herend is known for its porcelain factory, and while it was closed that day, many of the shops carried its wares. Who knew? NOTE: Eclipse trips have always had these serendipitous moments that just add spice.

Back at the restaurant Viktória, preparations began in earnest. I had a video camera on a tripod and staked out a prime piece of real estate in the field. The rest of the group that was documenting the event did likewise. Those simply viewing the event grabbed one of the many chairs outside the restaurant. We had a truly wonderful time.

I should mention here that the ladies in our group were quite appreciative of the fine restroom facilities that the restaurant had. That is definitely a factor to consider when looking for a viewing site.

As the partial phase of the eclipse (first contact and on) advanced, a few photos were taken. Some high, thin clouds persisted throughout the eclipse but never really interfered with viewing it. As the tiny crescent of the sun struggled against the encroaching moon, a cheer went up from everyone. The diamond ring lasted several seconds, and then there was darkness. Once more, the blackest of holes in the sky was surrounded by the pearly coronal light. The unhappy Hungarian chaperone now understood what all the fuss was about. She was absolutely amazed and not afraid to say so. The town streetlights came on, and birds went to roost at the noonday dark. The unnatural light gave the surroundings a very peculiar appearance.

As the moon receded and the sun once more dominated the sky, the owner of the restaurant broke out bottles of champagne for all as we toasted the eclipse. This would definitely go down as one of the most memorable days in my life. What had started out with some uncertainty had blossomed into a trip everyone would remember.

The next three total solar eclipses, (2001, 2002, and 2003) were in southern Africa and the Indian Ocean, so we decided to not attempt to see them. The years 2004 and 2005 had no TSEs, so we looked to 2006 for our next eclipse adventure.

Figure 11. The restaurant and hotel Viktória from where we viewed the TSE of August 11, 1999.

Figure 12. A salt shaker lid served as an impromptu solar projection device.

Figure 13. The street lights of Herend came on as the total eclipse approached.

Figure 14. Five ladies in the group enjoyed a bit of camaraderie by wearing identical, locally purchased, T-shirts.

8

Libya

On March 29, 2006, a TSE was predicted to travel through northwest Africa, turning slightly east as it emerged from the Libyan coast and crossing the Mediterranean into Turkey. I had read that traveling in Africa is usually more difficult than in Turkey, so I began researching trips to Turkey.

I had settled on a coastal location as a viewing site when I came across a brochure for an eclipse cruise that piqued my interest. It advertised viewing the eclipse from the Sahara Desert in Libya. Now, the years 2005 and 2006 offered a brief window when the US State Department allowed Americans to travel to Libya, and since the duration of totality was greater there than in Turkey, I figured I should try it. (Bonnie and I thoroughly enjoy cruising.) A group consisting of my wife and me, my brother and his wife, a friend of theirs, a former student, and three teachers arrived at Genoa, Italy, to board the MSC *Sinfonia* for a twelve-day Mediterranean cruise, including stops at places I had only read about: Naples, Syracuse, Alexandria, Tobruk (I had actually never read about, indeed, never heard of Tobruk), Tripoli, Malta, and Salerno. It was over my school's spring break, so I only missed a week of school. Man, was it worth it.

On this trip, I was not in charge. I did not have to arrange anything for the group. I was only a participant, and it felt good. There were over three thousand passengers on board—all there for the same reason as I: to see the eclipse. Once again I was in astronomy-nerd heaven. There were daily seminars on eclipses, with several big-name astronomers sharing their expertise. The ship was equipped with stabilizers; even at sea we were able to peer through the solar telescopes that were set up on the deck.

This was my brother's first eclipse, and I recall him making a point of telling me about his discussions with people on the ship who knew how many total minutes they had been in the moon's umbra (the full shadow) in the many eclipses they had viewed. I thought, "Yeah, I know that. No big deal." (It's now fourteen minutes, twenty-eight seconds.)

The weather was great for the entire trip. Spring in the Mediterranean was delightful. I did many firsts (and lasts), like ride a camel, see the pyramids, travel on a funicular, and enjoy the view of *Mare Nostrum* (the Roman name for the Mediterranean Sea) from a remarkably well-preserved ruin in Libya, Leptis Magna.

The day before E-day, we pulled in to Tobruk. The *Sinfonia* had never called on that port before. In fact, no ship in the MSC line had ever called on Tobruk, so this was a big deal. We watched from the railing as dignitaries arrived at the gangway and a ceremony marking the arrival and new relationship unfolded below us. Children brought flowers, and a great show of welcome and hospitality was made. I should mention that, in a somewhat ironic turn of events, shortly before our arrival at this port, the ship's officers informed all the passengers that Libya had added a one-hundred-dollar-per-person entry fee into the country, and as a courtesy, that charge would be added to our room bill.

Early the next morning, we looked out on the pier and saw fifty or more buses there to drive us two hours into the Sahara Desert to a site that had been set up for viewing the eclipse. One problem: there was fog in every direction. We boarded the buses in fog and drove through the city and into the country in fog. We spent the next hour trying to convince ourselves that the fog would not last and the sun would shortly burn it off. Well, it took longer than I had hoped, but it did, in fact, clear up.

The string of buses turned off the highway and trekked up a slight incline to a plateau. There was a tent city set up with vendors with all kinds of goods, along with dozens of porta-potties. We each grabbed a chair at a table for our group of nine and picked a spot from among the several acres where people were spreading out. We were in the company of thousands but literally in the middle of nowhere. The horizon was clear in all directions. No sand dunes here. Nothing but flat, pebbly, hard-packed soil as far as the eye could see.

This was a big event for Libya. Seldom had so many foreigners been allowed to enter the country. A news crew with a camera on a cherry picker documented the throng. Libyan Boy Scouts entertained with dances and songs. Merchants hawked their wares as the passengers filled the booths.

We set up our gear. I had brought two video cameras, one a three CCD affair, and my brother had a digital single-lens reflex camera with a telephoto lens. That took a little while. Then we walked around to inspect other setups. Many put our own efforts to shame. To say that there was an interesting juxtaposition of raw nature and high-tech telescopes and cameras is an understatement. But all that faded into the background with first contact. The sky was totally devoid of clouds. Only the sun and the silhouetted moon existed. As the moon advanced across the sun, the color of the sky became as rich and varied as I have ever seen it, and as the moon's shadow raced across the ground from the southwest and engulfed our temporary town, the rich colors of dusk washed over the entire horizon. Beautiful yellows, oranges, and reds were visible no matter which way you turned. And the crowd roared its approval and cheered encouragement for the moon to cover the last bit of the photosphere. For a few, precious seconds, a clear, crisp diamond ring adorned the sky. More cheers exploded as the last bit was covered.

And then there was the corona: that awesome black circle—a hole in the sky—was surrounded by gorgeous streamers of pearly light. Some of those extended three, four, and more solar diameters away from the black hole. Man, what an amazing sight—jaw-dropping, mind-blowing, breathtaking. Nearby, the planets and bright stars appeared. In the middle of the day! (Check out figure 15.) A hush fell over the spectators of this grandest of celestial events. Cameras snapped, and videos rolled as time seemed to stand still and we soaked up the rare ambiance of this syzygy. This was a moment worth savoring. But time moves on, and while this eclipse lasted for over four minutes, with the crowd roaring its encouragement, the moon began, all too soon, the lengthy reveal of the sun and allowed its return to its normal, dazzling self.

As the moon receded from the sun, the cruise-line staff provided a box lunch. While munching on a sandwich and fruit, the reality of viewing the eclipse from such a grand but desolate location began to sink in. It was doubtful that these circumstances would ever be repeated, so if I have said it before, I will say it again: this was a once-in-a-lifetime experience.

The remainder of the cruise was memorable in so many ways that the entire experience could be called grand. Welcoming ports combined with gorgeous weather and exquisite accommodations were a winning combination. As we made our way home, thoughts turned to the next trip, this one to a different part of the world: China.

I hoped that the eclipse of 2009 would provide as dramatic an experience as the Libyan one had, but it was not to be. Health issues forced Bonnie and me to relent and make the decision to stay home. As time passed, it became clear that the next opportunity for us would be the American eclipse of August 21, 2017. I knew that this time I could get more than just a few close friends and relatives to go see it. I have a base of astronomy, physics, general-science, computer-science, and math students that I have taught over the years who would benefit from and enjoy seeing this eclipse. Not to mention their family members, too. And the best part was, they wouldn't have to travel overseas. A simple drive would do.

What Am I Really Trying to Say?

Using myself as an example, I am trying to show you just what seeing a TSE can mean in your life. I want you to enjoy seeing and experiencing the sky, at day or night. I want you to understand and appreciate just how large, complex, and yet accessible is this grand thing we call the universe. In school most of the time, we focus on the details and miss the big picture. Not that the details are unimportant. They are, but they are also only part of the whole story.

I am an astronomy educator because of the early events that, while individually had little effect, collectively pushed me in the direction that ultimately brought me to the field I love and enjoy so much. That plus the solid shove provided by viewing a total solar eclipse. The reason I can relate all these stories of remembrance of all things ecliptic is because a few people at various times showed and explained things to me that caused me to want more, sparking the flame of curiosity within me and altering the course of my life.

I am a clear example of what can happen when you give a child, young person, teenager, or young adult an experience that is inspiring and impressive. The total solar eclipse of August 21, 2017, is sure to give millions a rare experience that will inspire and fill them with awe and most certainly yield fruit in the altered life paths that result.

Whether it is a career path to or lifelong interest in astronomy matters not. If they decide neither, at least they will have had the amazing experience of a total solar eclipse.

I hope that, by sharing my eclipse memories, I have passed along the desire to see one for yourself. So that's why I say:

1. **Go see the eclipse, and take a kid with you.**
2. **Never look at the sun without proper eye protection.**

Libya 2006

Figure 15. This image of the corona is a combination of eight images captured from a video of the TSE. The individual frames were averaged and processed to reduce the effect of overexposure of the inner corona when trying to image the outer corona. Getting good photographs of a TSE is not easy. Notice the ray structure going left and right. Each TSE has its own unique pattern of rays.

PART THREE
PLANNING FOR THE ECLIPSE

9

Why Go See *This* Eclipse?

So now let's get to the heart of the matter. Why should you go see this eclipse? What is the big deal? What is so great about this eclipse that you should take time off to travel, maybe hundreds or thousands of miles, to see it, especially when it only lasts a little over two minutes? Those are great questions that deserve answers. But to understand the answers, we need a bit more background.

I have been involved in astronomy education for over forty years. In the past several years, I have been regaling my students with stories, videos, and photos of total solar eclipses. I have traveled to see five of them—the last three at great distances because that's where they were. Unfortunately, America has been going through a bit of a dry spell when it comes to these total solar eclipses.

You see, even though TSEs occur almost every year somewhere in the world, they are rare at any given location. Even considering a large region like the United States, they are hardly common. Since 1900 (over a century ago), there have been twelve TSEs that have touched at least a portion of the fifty states (sometimes a very small part, at that). That's it. Twelve. In more than 110 years.

The last one to touch any of the fifty states was in 1991 in Hawaii. It occurred there just after sunrise. That one was also visible in Mexico, and many Americans traveled to see it. It was one of the longest-duration eclipses of the twentieth century, lasting just over seven minutes and well worth the trip. But it wasn't visible from the mainland. The last TSE visible from the forty-eight states was in February 1979. That one was obviously during the winter and traveled from Washington and Oregon to Idaho, Montana, and North Dakota, before heading north into Canada.

The last one visible from east of the Mississippi was in 1970. That one traveled north through Mexico, crossed the Gulf of Mexico, entered the United States in the Florida Big Bend area, and then traveled north through coastal Georgia, South and North Carolina, and Tidewater Virginia, where I saw it, before it passed into the Atlantic Ocean, making landfall again in Nova Scotia and Newfoundland. By the time the 2017 TSE occurs, it will have been forty-seven years since that last one seen in the east.

Every eclipse has a point in its path where the duration of the eclipse is the greatest for that particular eclipse. That is, the eclipse lasts longer there than anywhere else on its path. In *no case* did any of those twelve that touched the United States since 1900 have its point of greatest duration in the United States. (Granted, the eclipse of September 10, 1923, did come close.) So while we could have seen those eclipses from American soil, they were highest in the sky and of greatest duration outside the United States.

But why the emphasis on this particular TSE? After all, what you get for all your planning, organizing, packing, traveling, and setting up is just a few minutes of totality, even at the point of greatest eclipse. (Actually, for this eclipse, the maximum duration is a mere two minutes and forty seconds.) You might think you would prefer to stay at home and watch the eclipse on TV and not have to put up with the hassles inherent in traveling. Well, I'm here to tell you that seeing the eclipse on TV is nowhere near the same as being there and *experiencing* the eclipse. It is simply not possible for any camera and TV to faithfully reproduce the event. Full appreciation of a TSE only comes from personal presence. Don't do the television thing, except as a last resort.

On August 21, 2017, it will have been ninety-nine years (that's right, nearly a century) since the last total solar eclipse crossed the width of the forty-eight contiguous continental states. But even that eclipse failed to do two things that the 2017 eclipse will do: touch only US soil and have its point of greatest eclipse on US soil. If anyone in the whole wide world wants to see this eclipse from land, they have to come to the United States. (This is not a nationalistic statement, just geographic fact.)

Let's take a closer look at these two continent-crossing, ninety-nine-year-apart eclipses. The eclipse of June 8, 1918, started in the Pacific Ocean, just south of the main islands of Japan. The path of totality touched a couple of small islands off Japan and then crossed the expanse of the North Pacific, skirting the Aleutian Islands. The point of greatest eclipse occurred over the North Pacific Ocean. The TSE next made landfall in southern Washington, its sixty-mile-wide path passing between Olympia and Portland, Oregon. The path headed east and a little south, missing many cities. But it did pass through Denver, Colorado, then a city of a quarter-million people and the largest city in the path of the eclipse. The path continued into the southeast, passing through Oklahoma, Arkansas, Mississippi, Alabama, and Florida. It exited the United States and ended in the northern Bahamas.

The 2017 eclipse will have a very similar path. But a few differences will make this eclipse even more special for people in the United States. (See figure 16.) It also will originate in the Pacific but closer to America. And, in contrast to the 1918 eclipse, the 2017 point of greatest eclipse will be over land, specifically in southern Illinois and southwestern Kentucky. The path on land will be similar, starting in Oregon between Portland and Eugene. It will travel mostly east, slightly south, touching Idaho, Montana (barely), Wyoming, Nebraska, Kansas, Iowa (again, barely), Missouri, Illinois, Kentucky, Tennessee, North Carolina, Georgia, and lastly South Carolina. This is the American eclipse. That is, if anyone wants to see this eclipse from land, they have to be in the United States.

It is interesting to note that the path of the 2017 eclipse across the United States ranges from 60 to 72 miles wide and about 2,500 miles long, covering an area of about 162,000 square miles. Yet the largest city that it covers in its entirety is Nashville, Tennessee, the nation's twenty-fifth-largest city, according to the 2010 census. (See figure 17.) In the appendix there is a list of the cities in the path of the eclipse and those close to but not in the path.

Goals: My Challenge to You

Within two hundred miles of the centerline of the eclipse live an estimated sixty-five million Americans. What I would like to see happen during this eclipse has several parts, in order of increasing complexity and difficulty to achieve:

1. That every school-age child in the path of totality gets to see the eclipse.

This is eminently feasible. There may be a few exceptions, but otherwise, school-age children must (all right, *should*) be allowed to see this eclipse. It holds the greatest promise for giving our country a boost in science education.

2. That every person in the path of totality sees the eclipse.

Perhaps not quite as feasible, but still, a very high percentage of the adults living in the path of totality, even those working, should be given the two to three minutes required to view totality, if not the entire partial phase of the eclipse.

3. That every person within two hundred miles of the centerline sees the eclipse.

I especially want children to see the eclipse, but if they live outside the path of totality, common sense dictates that they must be driven/transported to a site inside the path by an adult. For this and the next group especially, I say, "Go see the eclipse, and take a kid (or two) with you."

4. That as many people as possible from the rest of the United States and the world see this TSE.

August 21 is still during summer vacation season for many people. I can think of no better way to spend part of your holiday than going to see a total solar eclipse. It may be that those who live in New England or Southern California have to plan a little more since their journey will be a major one, but trust me, it will be worth it. Those coming from overseas will, of course, have much to contend with in planning such a trip, but again, it will be worth it.

It is somewhat disconcerting to me that, in the recent past, people have forbidden and prevented children from seeing total solar eclipses because they had heard of the harm that can occur to the human eye. I will address that issue later and try to put to rest any fears that you may have about viewing the eclipse. After all, travel companies have lively businesses taking people into the path of totality to see the event. There must be a way to view it safely.

There are several reasons for my interest in this eclipse. As an astronomy educator, I feel my greatest joy comes from showing people something in the sky that they have not seen before and seeing and hearing their expressions of awe and wonder. In studying the history of astronomy, I have read of many people who saw something in the heavens—a supernova, a large bright comet, or a total solar eclipse—that inspired them and changed the course of their lives. Most of the time, these sights were seen while they were with children.

A total solar eclipse is a great teachable moment. Teachers can use eclipses to show how the laws of motion and the mathematics of orbital motion can predict the occurrence of such events. The use of pinhole cameras and telescopes or binoculars to observe an eclipse can lead to an understanding of the optics of these devices. The rise and fall of environmental light levels during an eclipse illustrate the principles of radiometry and photometry. Even students taking biology can observe the associated behavior of plants and animals. It is also an opportunity for school-age children to contribute actively to scientific research. Making observations of and recording contact timings at different locations along the eclipse path are useful in refining our knowledge of the orbital motions of the moon and earth, and sketches and photographs of the solar corona can be used to build a three-dimensional picture of the sun's extended atmosphere during the eclipse.

Nearby supernovas are spectacular but extremely rare, and their occurrence is not easily predicted. Bright comets are also unpredictable and, while not as rare as supernovae, rare enough. Both are best viewed at night. But with increasing light pollution in cities and towns everywhere, they are not the grand sight they would be from a remote, dark location. A friend of mine who is a basketball coach told me of a tournament his school held a few years ago. The school was in the suburbs. A team from the inner city participated in the tournament. At one point, while going from the gym to the bus after the game, they marveled at lights in the sky and wondered what they were. They were stars. The city lights of home prevented them from seeing any stars. That just blew my mind. I have come to the realization that, for a large segment of the population, nighttime viewing of celestial wonders is going to be severely limited from now on. That makes TSEs even more important, since they occur during the day and can't be missed.

Total solar eclipses at any given location on earth are also rare, occurring about every three hundred to four hundred years. But they are *very* predictable. Today we know with almost absolute precision when and where the next century's worth of eclipses will occur. So this eclipse is coming. We know when and we know where. Because TSEs are rare in any given location, that means they usually occur somewhere else in the world and require travel, sometimes over great distances, to see them. As I have said, I have seen one from America, one from the Caribbean, one from Hungary, and one from the Sahara Desert in Libya. I went to see one in Canada, but the weather did not cooperate, and we saw clouds instead. Real eclipse chasers think nothing of traveling twenty-four hours to Siberia or Sulawesi, Indonesia, to see a few precious minutes of totality.

This brings me back to reality. The eclipse will occur on August 21, 2017, but if clouds are in the way, even if you are in the path of totality, you will not be able to see it. So there is some risk and uncertainty in traveling to see the eclipse. But even if there is cloud cover, you will experience the eclipse. It will get noticeably darker, cooler, and eerier. The quality of light in the sky will change and be different from anything you have experienced before. So it will not be a total loss. Disappointing, for sure, but not a total loss. That's a risk I am willing to take.

We can study historical weather data, such as humidity and average cloud cover, but on the day of the event, history does not count for much. Only that day, weather does. Still, the prospect of seeing an eclipse has sent people traveling thousands of miles. For this one, it is in our own backyard, so to speak.

Some Americans, granted, will have to travel quite a bit to get in the path. But even the farthest distance is doable. People from San Diego, California, and Brownsville, Texas, will have to travel a thousand miles—eight hundred for people from Maine and New Hampshire. No one in the forty-eight states must travel farther than this to see the eclipse. Traveling that far is a major trip and will require much preparation and effort. But it beats having to fly to Europe or Australia to see a TSE. And the fact that it occurs during what is, for many, summer vacation, makes the prospects for full campgrounds and hotels in the path of totality almost a certainty. So start making plans. For a list of towns inside the path of totality of the 2017 total solar eclipse, see appendix III.

One final reason to go see the eclipse of August 21, 2017

Americans, as a people, seem to desire unique and out-of-the-ordinary experiences. They enjoy unusual, rare, and spectacular events. Millions visit our great natural wonders: the Grand Canyon, Yellowstone, the Smoky Mountains, Acadia, and Niagara Falls, to name a few. They get a thrill from experiences that challenge them. They get a rush by doing the different and unusual. Well, if you don't get anything else from this chapter, please get this: A total solar eclipse qualifies on all these accounts. It is rare, it is unusual, and it is exciting and truly spectacular. You may not think of yourself as a science type. That's OK. You don't have to be to enjoy an eclipse. So don't let that keep you from going to see this one and missing out on one of the most memorable events of your life. Remember:

1. **Never look at the sun without proper eye protection.**
2. **Go see the eclipse, and take a kid with you.**

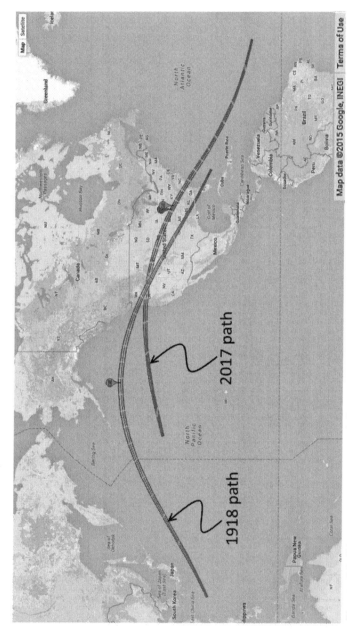

Figure 16. Direct comparison of the paths of the TSEs of 1918 and 2017. Map courtesy of Google Maps. Eclipse path by Xavier M. Jubier using data from Fred Espenak, NASA.

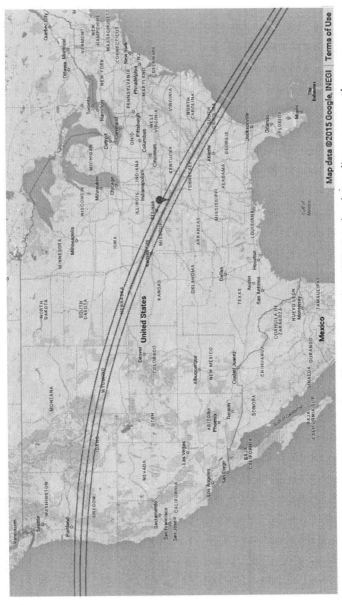

Figure 17. The portion of the path of the August 17, 2017, total solar eclipse across the United States. Map by Google Earth. Eclipse path by Xavier M. Jubier using data from Fred Espenak, NASA.

10

Logistics

OK, so you have decided to go see this eclipse. That's great. Congratulations. Now what? What do you do? Where do you go? What do you need? So many questions.

Well, at the simplest level, all you have to do is find a place to view the eclipse and stay there for the few hours. Afterward, celebrate a little and go home. Of course, the reality is usually a bit more complicated. So let's look at several possible scenarios for viewing the eclipse. Hopefully one of them will fit your situation closely and within reason. These scenarios mostly depend on where you live and how much time you have to plan your trip. Remember that the eclipse is on a Monday.

Scenario 1: *You are already in the path of totality.* I estimate that ten million people live inside the path. This is only about 3 percent of the US population but is a sizable number nonetheless. (See appendix III for a list of towns inside the path of totality.)

This is probably the easiest scenario to deal with. You will not have to travel much, if at all, on eclipse day. You can view the eclipse from your front yard, a park, or any open place where the sun can be seen. The only exception to this is if you live right on the edge of the path. There are two main reasons you might not want to be on the outer edge of the eclipse: totality will be short (thirty seconds or less), and there might not be totality at all. That's right; our ability to predict the edge of the shadow of the moon is less than perfect because the moon is not a smooth sphere but has mountains, craters, and valleys. That means that the shadow has wiggles in it that are difficult to predict. So if you live right on the edge of the path, you will certainly want to move to a place that is a few miles closer to the centerline to guarantee that you will have some totality.

And then there are a couple of reasons why you might choose to travel on eclipse day. One is to meet with friends and family for a memorable eclipse party. That is an excellent idea. The other is if you are one of the many who will volunteer/work at a community center/church/school/business or other location that has agreed to welcome visitors and provide a safe place for them to view the eclipse. I sincerely hope that hundreds, if not thousands, of sites are set up to do just that for this event. (More on that in chapter 12, "This Calls for a Celebration.") But that leads to the one certainty about this eclipse: on eclipse day, millions of people, who would normally be somewhere else, will be in the path of totality, possibly near where you are. Not just a few people but thousands, tens of thousands, or even hundreds of thousands—maybe even millions—of people will pour into the eclipse path to view this grand spectacle of the heavens.

CAUTION: If you live just outside the path of totality, perhaps a mile or two, do not think that you will see the total eclipse. You won't. You will see a very skinny sun mostly hidden by the moon, but you will *not* see the corona. You will *not* see the diamond ring. You will *not* see any of what makes this event so special. *Please* move a few miles so you are certain you are in the path of totality. You will thank me later.

Scenario 2: *You are not in the path of totality but you are within two hundred miles or so of the centerline.* Seeing the eclipse could possibly be a day trip.

If you have the time, scout out potential sites for viewing the eclipse several days or weeks in advance. This can be done by taking a trip to a town in the path, looking for a public place like a park or community center, and paying close attention to conveniences, such as parking availability and the location of public restrooms. An open field will guarantee that no trees or buildings will block the sun on eclipse day. Some communities may actually take advantage of their location in the path of totality and have special facilities set up for visitors that day.

The point is that, if you have the time, a little preparation can go a long way toward making this eclipse a positive and enjoyable event for you and your friends and family.

A typical timetable on eclipse day involves getting an early start. For instance, in Oregon, you should leave home for the viewing site no later than 5:00 a.m. In Tennessee it would be 7:00 a.m. (More details on the timing of the eclipse coming up later in this chapter.) At any given site, the entire eclipse lasts for a little less than three hours, so you could be home by late afternoon or early evening. Of course, considering that there may be many others traveling on eclipse day, you should allow for delays.

Scenario 3: *You are not in the path of totality and more than two hundred miles from the centerline.* This will likely require an overnight stay in the eclipse path. If you have friends or relatives you can stay with, great. I suggest that you be extra nice to them between now and the eclipse. Just saying. If you are going to camp or use a motel, you will almost certainly need to make arrangements and reservations well in advance.

This eclipse will be widely advertised, and many people are going to travel, some great distances, to experience it. This includes Americans as well as those coming from other countries. Remember August is vacation time for a lot of people, and seeing the eclipse will make a visit to America just that much more attractive.

Scenario 4: *You are not in the path of totality and had not planned on going to see the eclipse, but a change of circumstances (or heart) has prompted you to try to see the eclipse at the last minute.* If this is you, you will just have to wing it. There will probably be apps and websites that show places to go (be sure to visit www.goseetheeclipse.com for updated info) with parking, restrooms, and food available. I'm certain that you will not be alone. Many people can't control their schedule, and free time may become available at the last minute. But do you know what? That's OK. There is no shame in that. My wife and I have done some crazy things, like drive all night, to see the space shuttle launch or something similar. Where there's a will…The important thing is to *go*.

11

Eye Safety
and Viewing the Eclipse

Equipment/Material You Might Need

The truth is you don't need *any* equipment to view the total eclipse. Your naked eye does just fine when it comes to watching totality. That being said, don't let your guard down. The total phase of the 2017 eclipse lasts, at the most, two minutes and forty seconds. The partial phase, on the other hand, lasts an hour before and the same amount after totality. Do not look at the sun during the partial eclipse, unless you have proper viewing equipment. I REPEAT: *Do NOT look at the sun at any time, other than totality, without proper viewing equipment. Staring at the sun could cause permanent damage to your eye. It isn't worth the risk.*

So, what might that equipment be? The methods for viewing the sun fall into one of two categories: projecting the image of the sun and viewing the sun through a filter to reduce the intensity of the sun's light to a safe level. I mentioned this earlier in the book, but it is worth repeating. We will deal with projection methods first.

Solar Projection

This is for the partial phases. Projecting the sun's image is generally considered the safest method of viewing it since you do not have to face the sun at all, so there is little chance of accidentally looking at the bright sun.

Projection requires an opening of some kind for light to pass through. In this case, the smaller the opening, the better. The simplest type is just a pinhole—no lens, no complicated apparatus, just a tiny hole. Nature may even provide one for you. If the sun is out and there is a tree nearby, walk underneath it and look at the ground (look for a fairly smooth area under the tree). The many irregularly shaped leaves overlap, leaving little gaps that, in many cases, act just like little holes. When we look down on the ground under a tree most days, the many lights are round, which happens to be the same shape of the opening (or nearly so), as well as the same shape as the sun, so we don't think much about it. Well, the image on the ground is that of the sun, not the round opening. Only during an eclipse do the little images start to appear different, gradually becoming crescent shaped as the partial phase proceeds.

If there are no nearby trees or you want to try making a projection device, there are a few things I have tried. I give detailed directions on how to build and safely use these in Part IV: Activities.

Some refracting telescopes have projection screens and adapters to see a larger (that is, magnified) image of the sun so that several people at a time can see it. If you own or buy one, please read the instructions that come with it on how to attach and use the components safely. There are simply too many variations for me to give detailed information on each. *Please view the eclipse safely.*

Solar Filtering

In order to safely look at the sun (as opposed to looking at a *projection* of the sun), the amount of light coming from it must be greatly reduced. This is discussed extensively in an article by B. Ralph Chou on a NASA website dealing with eye safety during solar eclipses. A portion is included here (used by permission of Dr. B. Ralph Chou).

The sun can only be viewed directly when filters specially designed to protect the eyes are used. Most such filters have a thin layer of chromium alloy or aluminum deposited on their surfaces that attenuates both visible and near-infrared radiation. A safe solar filter should transmit less than 0.003% (density~4.5) of visible light (380 to 780 nm) and no more than 0.5% (density~2.3) of the near-infrared radiation (780 to 1400 nm). [footnote omitted]

It continues:

One of the most widely available filters for safe solar viewing is shade number 14 welder's glass, which can be obtained from welding supply outlets. A popular inexpensive alternative is aluminized Mylar manufactured specifically for solar observation. (**"Space blankets" and aluminized Mylar used in gardening are not suitable for this purpose!** [emphasis added]) Unlike the welding glass, Mylar can be cut to fit any viewing device and doesn't break when dropped. Many experienced solar observers use one or two layers of black-and-white film that has been fully exposed to light and developed to maximum density. The metallic silver contained in the film emulsion is the protective filter. Some of the newer black and white films use dyes instead of silver, and these are unsafe. Black-and-white negatives with images on it (e.g., medical X-rays) are also not suitable. More recently, solar observers have used floppy disks and compact disks (both CDs and CD-ROMs) as protective filters by covering the central openings and looking through the disk media. However, the optical quality of the solar image formed by a floppy disk or CD is relatively poor compared to Mylar or welder's glass. Some CDs are made with very thin aluminum coatings, which are not safe—if you can see through the CD in normal room lighting, don't use it! No filter should be used with an optical device (e.g. binoculars, telescope, camera) unless it has been specifically designed for that purpose and is mounted at the front end (i.e., end towards the sun). Some sources of solar filters are listed in the following section.

Unsafe filters include all color film, black-and-white film that contains no silver, photographic negatives with images on them (X-rays and snapshots), smoked glass, sunglasses (single or multiple pairs), photographic neutral-density filters, and polarizing filters. Most of these transmit high levels of invisible infrared radiation, which can cause a thermal retinal burn...The fact that the sun appears dim, or that you feel no discomfort when looking at the sun through the filter, is no guarantee that your eyes are safe. Solar filters designed to thread into eyepieces that are often provided with inexpensive telescopes are also unsafe. These glass filters can crack unexpectedly from overheating when the telescope is pointed at the sun, and retinal damage can occur faster than the observer can move the eye from the eyepiece. Avoid unnecessary risks.

For more information, see NASA's web page "Eye Safety During Solar Eclipses" at http://eclipse.gsfc.nasa.gov/SEhelp/safety.html.

So you can see that this is no time to be careless. Eyesight is precious. Use a number 14 welder's glass or a filter specifically designed for viewing the sun safely. Since the 1998 TSE, I have used eclipse shades or similar glasses to view the partial phases of the eclipses. One source is http://www.rainbowsymphony.com/eclipse-glasses.html. You can find additional sources at this book's companion website, http://goseetheeclipse.com.

Where and When to Go

The eclipse travels from west to east. The eclipse centerline touches twelve states. What follows are tables showing the circumstances of the eclipse as it first touches each of those states on its western side. I will attempt to explain the relevant terms.

There are four *events* to an eclipse: first contact, second contact, third contact, and fourth contact. I will not go into a detailed explanation of these. All a novice needs to know is that the amount of time between first and second contacts and between third and fourth contacts is long—over an hour each. These two time periods comprise the partial eclipse. The time between second and third contacts, however, is less than three minutes. That period is totality. That is the real show.

At first contact, the moon appears to just touch the sun. It is difficult to observe, so don't try, unless you have the right equipment. For the next hour and fifteen minutes or so, the moon covers more and more of the sun. It is barely discernable at first, looking as though a bite has been taken out of the sun. Only after the sun's diameter is more than 50 percent covered does the average observer usually notice the eclipse. Gradually the pace appears to pick up. The moon doesn't move any faster; it just seems to. Changes begin to occur. Once the sun is more than 75 percent covered, the sky begins to darken. As more of the sun disappears from view, the sky takes on a dim color, somewhat like dusk, but not quite. Projections of the sun clearly show a crescent, which gets smaller and smaller by the minute. Even at this point, *do not look at the sun*. If you have solar shades designed for viewing an eclipse, make sure that you have them in place when looking at the sun. THIS IS NOT A GAME. DO NOT LOOK AT THE SUN WITHOUT PROPER EYE PROTECTION.

At second contact, the total eclipse begins, kicked off by a glorious diamond ring. Take off any filters you have and look up. Until third contact (and a second diamond ring), you will have a front-row seat at one of nature's grandest displays. It is OK to cheer and clap. Or stare in awe and wonder. Or both. It's all good.

How will you know when third contact happens? Remember, the moon is continuously, if slowly, moving. It covered the sun by moving in front of it and now will gradually get out of the sun's way. The first little bit of the sun that peeks out will produce the second diamond ring of the event. That is third contact. Make sure your filters are back in place. That little bit of sun will gradually grow back into a crescent, and the sky will gradually, in the course of an hour of so, revert back to its original brightness and color. The sun will be fully uncovered, and the eclipse is over.

Back to the tables. Table 1 contains the exact times of the four contacts for where the eclipse path enters Missouri (from the west, remember). How do you read the table? It certainly contains a lot of information, mostly numbers, which can become a blur. But don't let that scare you away.

Table 1. Center Line Circumstances when the sun's shadow enters Missouri from the west.

Eclipse Events in Missouri	Time (CDT)	Alt	Azi
Start of partial eclipse (C1) :	11:40:36.6	54.2°	134.0°
Start of total eclipse (C2) :	13:06:22.8	61.9°	171.6°
End of total eclipse (C3) :	13:09:01.0	61.9°	173.0°
End of partial eclipse (C4) :	14:34:30.8	57.9°	214.7°

The rest of the state time tables are in Appendix II.

First, let's look at the abbreviations. *C1* means first contact; *C2*, second contact; *C3*, third; and *C4*, fourth. All times are in twenty-four-hour clock notation. *CDT* stands for central daylight time and is used for the states in that time zone, in this case, Missouri. (If a state is divided by two different time zones, the table would refer to its western time zone.) *Alt* and *Azi* refer to coordinates to help locate the sun in the sky. *Azi* refers to *azimuth*, or the heading along the horizon. Due north is 0 degrees azimuth, due east is 90 degrees, south 180 degrees, and west 270 degrees. *Alt* refers to *altitude* in degrees above the horizon. Values for altitude range from 0 degrees (at the horizon) to 90 degrees straight overhead.

Let's take an example. Suppose you are going to view the eclipse in St. Joseph, Missouri. The chart for Missouri (the western edge next to Nebraska) says the eclipse starts (C1) at 11:40:36.6 CDT. This marks the start of the partial eclipse. It lasts for one hour and twenty-six minutes until C2 at 13:06:22.8 CDT. That marks the start of totality. C3 (the end of totality) is at 13:09:01.0 CDT. The second partial phase then lasts until 14:34:30.8 CDT (C4). At that point the eclipse is officially over for St. Joseph, Missouri.

Approximations for locations between state lines can be made by figuring how far you will be from the western edge of your state. Now, the shadow of the moon travels very fast. To give you an idea of how fast, it travels from the coast of Oregon to the Idaho–Oregon border (320 miles) in nine minutes. So these tables give an approximate idea of the time the eclipse occurs at your location. See appendix II for the other states.

I need to make a couple of points here:

I mentioned that the shadow of the moon travels quickly. In looking at the times and taking into account the different time zones, it takes the moon's shadow about one hour and twenty-five minutes to travel from the coast of Oregon to the Atlantic Ocean at South Carolina, a distance of 2,500 miles. That works out to about 1,750 miles per hour (2,900 kilometers per hour). Now that is fast!

Since August is still summer, the sun is pretty high in the sky during totality, over 60 degrees in all but the first few states. This offers the best chance for visibility, even if it is partly cloudy that day. Clouds are often thickest along the horizon and thinnest higher up in the sky.

Please check the blog at goseetheeclipse.com or the Facebook page Go See The Eclipse for more tips and suggestions for experiencing the total solar eclipse of August 21, 2017.

And again:

1. Never look at the sun without proper eye protection.
2. Go see the eclipse, and take a kid with you.

12

This Calls for a Celebration

At the heart of the eclipse are two main ideas. First, if you want to see the eclipse, get somewhere in the path of totality, on time, otherwise you will miss it, and, second, after the eclipse, be prepared to celebrate.

Let's compare the eclipse to a birthday party. The excitement that my own granddaughter experiences as her birthday approaches can reach epic proportions. A similar feeling occurs with a total solar eclipse. It may be weeks, months, or years away, but you know that it is coming. Most people would say that the best part of their birthday party is opening up the presents. After all, who doesn't enjoy receiving a gift? I want you to think of this journey to view the TSE as a party, with the *gift* being the eclipsed sun given to each one who attends. Witnessing one of nature's grandest spectacles is cause for a party if ever I heard of one. And you will not want to miss it.

In some ways, I'm a slow learner, and the following list comes from my five attempts to see a TSE. You want to get it right the first time, because these opportunities don't come along every week, month, year, or decade. I hope these eclipse party tips will help you achieve the goal of experiencing the eclipse and doing so successfully.

Tip 1: Provide viewing devices to be sure each person's eyes are safe. After the eclipse, keep the devices to look at the sun and remind yourself of the wonderful experience you had.

Tip 2: Invite as many people as possible to see the eclipse with you. "The more, the merrier" is certainly true here. And please include plenty of school-age children.

Tip 3: Make sure you have a main viewing site and an alternate viewing site planned and available. Then find out the start and end times of the eclipse at these locations.

Tip 4: Check the weather forecast frequently for several days before the eclipse. If clouds threaten the morning of the eclipse, be prepared to quickly change to your alternate viewing site.

Tip 5: Depart from home in plenty of time to arrive at your viewing site before the eclipse starts. If it is just a day trip, be certain to leave early enough to allow for travel delays. If you can afford the time, arriving a day or two ahead of the eclipse is not unreasonable.

Tip 6: Bring observing equipment (telescopes, binoculars, viewing glasses), chairs, food, beverages, cameras with extra batteries, and noisemakers (it will be a celebration).

Tip 7: As totality approaches, cheer and yell. Really whoop it up. After all, what's a great party without audible signs of celebration? You are about to open your eclipse gift. Do the noisemaking early, because once totality is underway, you will be slack-jawed with awe. I kid you not.

PART FOUR
ACTIVITIES

Make the Eclipse a Memorable Experience for You and the Children Around You

If there is one thing I have tried to communicate in this book, it is that I want you to see the eclipse and see it in a prepared, meaningful way. Since there is no way to get a sneak preview of a TSE (sorry, but there just isn't), you have to prepare by other means. Photos and videos of an eclipse can help, but they are both poor representations of the real thing. That being said, you should still prepare yourself as well as you can. (I was a Boy Scout, and their motto is "Be prepared.")

On the following pages are some activities—teachable moments you can do with your children. These are designed to improve your understanding of the causes and events surrounding the eclipse and will help you get the most out of this experience. These are optional but highly recommended.

1. Visualize the scale of both the size and separation of (distance between) the three objects involved in this event: the earth, the sun, and the moon.

For this activity you will need a marble (about one centimeter or half an inch in diameter), a BB or small ball bearing (about one quarter of the diameter of the marble), and the biggest beach ball you can find (about three feet in diameter). If no beach ball is available, get a piece of string ten feet long, and knot the ends together to make a closed loop. Next, find a football field. (If no football field is available, find any large field, and measure about four hundred feet.)

At one end of the field (the corner, actually) set the beach ball down or lay the string on the ground in a circular shape. That represents the sun. Now go all the way to the other corner of the other end of the field, including the end zones. Place the marble and BB about one foot apart, with the BB placed between the marble and beach ball/string circle. Now *that* is a good approximation of the size-and-distance scale of the sun-moon-earth system during the eclipse (and any other time, for that matter). Just move the BB around the marble in a circular path to mimic the monthly orbital motion of the moon around the earth.

For those of you who want numbers, the ratio of diameters of the three objects is approximately:

moon   = 1
earth   = 4
sun   = 400

The distance from the earth to the moon is thirty times the diameter of the earth.

The distance from the earth to the sun is four hundred times the distance from the earth to the moon.

Some of the concepts you can get from this exercise are:

- The solar system is large, and it is mostly space.
- We can see some things that are very far away.
- When we look at the sun and the moon, they appear to be about the same size. They aren't the same size, of course, but they look the same size. That is what allows us, on the earth, to see the moon completely cover the sun.

Whenever I do this activity, it blows my mind. To see these three objects at this scale with accurate distances is amazing, not to mention that, at this scale, the planet Jupiter would be the size of a softball nearly half a mile from the sun, and Pluto would be a BB three miles from the sun.

Let's examine this in a little more detail. Notice that the ratio of the diameter of the sun to the diameter of the moon (four hundred to one) is *the same* as the ratio of the earth-sun distance to the earth-moon distance (four hundred to one). So even though the sun is four hundred times farther away from the earth than the moon, it will appear to us on earth to be the same size as the moon because it is four hundred times bigger than the moon. One more important fact is that both the sun and the moon are close to being spherical.

If all that seems a bit confusing, just remember this: since the sun and moon appear to be the same size in our sky, the moon can totally cover (or eclipse) the sun and hide it from view. As far as we know, no other planet in the solar system has a moon that can do that. Ours is the only planet where you can see a total eclipse of the sun. So take advantage of it.

2. Make your own eclipse.

Version 1: You may recall that I mentioned this in chapter one. Let me expand on it a bit. You will need a quarter and a dime. Hold the quarter in one hand between the thumb and first finger so that you can see the circle, not the edge. Extend your arm full length. In your other hand, hold the dime in the same manner but closer to your eye. Close the other eye. Gradually move the dime toward or away from your open eye until it appears to completely cover the quarter. Move the dime away until it just barely covers the quarter. Notice that if you move the dime any farther away, it no longer appears large enough to completely cover the quarter. In the same way, if our moon were farther away from the earth, it could not fully cover (eclipse) the sun.

Version 2: You will need a flashlight and a quarter. Lay the flashlight on a shelf or table, pointed toward an open part of the room. Turn the room lights off. Stand several feet away, and look at the light. Close one eye. While looking at the light, hold the quarter between your thumb and first finger at arm's length. By adjusting the position and distance of the quarter from your open eye, you should be able to block the light from the flashlight. Notice that some light that is scattered by the flashlight is still visible, even when the light is completely covered. That is an eclipse.

3. Prepare to view the eclipse with proper eye protection.

This is *critical*. The sun is so bright that you should NEVER (YES, I'M SHOUTING), EVER look directly at it. There is a reason why it hurts to look at bright lights, the sun being a good example. The pain is your body's way of telling you to stop what you are doing. You might harm yourself. That being said, there are ways to look at the sun.

While I have touched on this earlier, eye safety is so important that I mention it here again. The proper viewing of the sun can be done in two primary ways:

1. filtering the bright light from the sun; and
2. projecting the image of the sun.

A Closer Look at Solar Filters

Filtering the sun's bright light is accomplished today with commercial filters. I don't even want to mention some of the methods used in the past to block sunlight, but a few are described in the articles referenced below. In the recent past, I have used cardboard viewing devices (they look like glasses) that have Mylar material coated with a thin layer of aluminum or similar material that allows only 1/100,000th of the sun's light through. They are designed for viewing the sun, can be used safely, and are inexpensive. They are available online but should be ordered well in advance of the eclipse to ensure delivery before eclipse day. The best case is when each person in your observing party has his or her own pair.

You must check each pair you get to verify that they have no holes or defects in them—*each time you use them*. Eye safety is *not* a game.

For more information on eclipse-viewing safety, visit http://eclipse.gsfc. nasa.gov/SEhelp/safety.html.

For an even more detailed examination of the topic, visit http://www. mreclipse.com/Special/filters.html.

There is so much information about projecting the sun's image that this next section is devoted to it.

What to Make and Take to the Eclipse

1. Index-card projection.

Projecting the image of the sun usually involves two objects: one with an opening and another to act as a screen. A simple setup comprises two three-by-five-inch note cards. On one of the cards, use a needle to poke a small hole. The smaller, the better. Then stand with your back to the sun. Using one hand, hold the card with the pinhole up by your shoulder. Take the second card (the *screen*) in the other hand, and hold it right next to the pinhole card. Gradually move the screen card down and away from the pinhole. You should see a bright round dot on the screen. Continue to move the screen away, keeping the bright dot near the center of the screen. The image you see is of the sun. The farther you move the screen away from the pinhole, the bigger the sun's image. If you move it too far away, the image will be so big that the image will be spread out to the point that it will be too faint to see easily. Do *not* be tempted to make the hole bigger to let more light through. That will only reduce the resolution of the image.

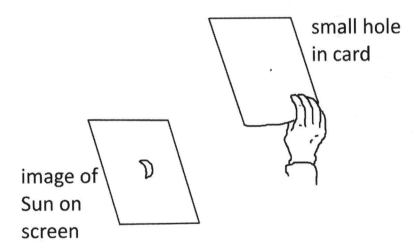

small hole in card

image of Sun on screen

Figure 18: Pinhole projection with note cards. This setup requires no lenses or expensive equipment.

If you happen to own a refracting telescope, the image of the sun improves, but the setup gets more complicated. You can use such a telescope to view the projected image of the sun, and it will be bigger, brighter, and better resolved than the note-card method—but at a cost. A telescope will require you to constantly change its position to keep the projected image on the screen. There are so many combinations and variations in configuration that I cannot cover them here. Rather, I am going to tell you to check your owner's manual for instructions on how to use your telescope to view the sun, whether it is with a filter or with projection.

Just remember: NEVER LOOK AT THE SUN WITHOUT PROPER PROTECTION. The risk is just not worth it.

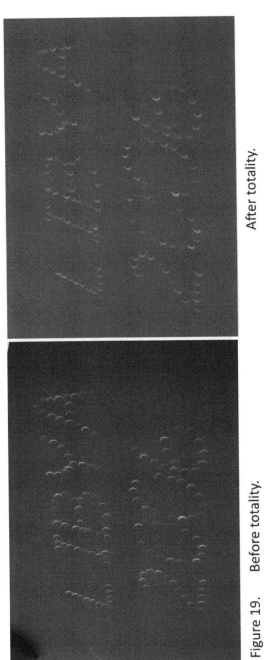

Figure 19.   Before totality.   After totality.

Notice the change in the orientation of the crescents.

2. Pinhole projection of the place and date.

Preparation: A few days before the eclipse, take a sheet of heavy paper/ card stock, and draw in big letters the place and date of the eclipse or other message of your choice. Using a pin, poke small holes equally spaced along the letters. Each pinhole is going to project a small image of the partially eclipsed sun onto another sheet of paper or other white surface.

Eclipse day: Make sure you have a way to hold both the card stock with the pinholes and a sheet of white paper or a tabletop surface on which to project the image. Before the eclipse begins, the pinholes will project as circles. As the partial eclipse begins, the images will be circles with bites taken out of them. Closer to totality, the images will be crescents. (See figure 19.) This will not work during totality. Look directly at the sun then. After totality, the images will be crescents on the opposite side of the little circles.

3. Shoe-box projection.

Another version of this device has been used for quite a while and involves a box. Before eclipse day, get a shoe or similar box. You should remove the lid. You don't need it for this activity, although you may want to keep it for transport. (See figures 20–25 for this section.) Cut an opening that is about one by one inch in one end of the box. An adult should do this. Get a piece of aluminum foil big enough to completely cover the opening you just made. Handling the foil very gently, tape it to the box, over the opening you just cut in the end. Using a sharp pin or needle, carefully punch a small, round hole in the foil. Generally, the smaller, the better. Make certain that the tiny hole is through the opening, not the box. Tape the edges of the foil completely. You don't want it to come loose. Tape a three-by-five-inch, white (unlined) note card or white sheet of paper cut to size on the opposite end of the shoe box (on the inside). That's your pinhole projector.

You can test it out ahead of time by going outside (assuming it is sunny). Stand with your back to the sun. Hold the box, with the foil pinhole end up, aimed at the sun. Light from the sun will come through the pinhole. Adjust the position of the box so that the round image of the sun lands on the note card on the side opposite the foil pinhole of the box. The little circle of light falling on the note card/paper is the sun. On eclipse day, as the partial phase progresses, you will notice the sun has a bite taken out of it by the moon.

This will also work with a postal mailing tube or any long container where you can make a pinhole at one end and put a note card on the other for the screen. You may have to cut some of the side of the tube away so that you can see inside it.

Projection devices usually allow two or three people to view the image simultaneously.

4. Other things to gather and take with you.

Thermometer to Show Temperature Change.

Get a thermometer (digital or traditional), and record the temperature at regular intervals (say, every ten or fifteen minutes). After the eclipse, you will have enough data to create a nice table and graph. What does the data tell you about the eclipse's effects?

Light Meter or Photo Resistor to Show Light Changes.

This activity is a little more complex. It depends on having a light meter or photo resistor. I might suggest using the one in a camera, but during the eclipse, my cameras are all busy taking photos of the eclipsing sun. Aim the light meter at the same spot in the sky (but not at the sun), and record light-level readings at regular intervals. Again, at the end, you will have enough data for a nice table and graph. What does the data tell you about the eclipse's effects?

Video Camera to Record Personal Impressions.

This can be done in many different ways. If you have a video camera running, just record your thoughts and observations as it runs.

Paper/Notebook and Pencil for Journaling.

During the partial phases, things are moving slowly enough that you can write your thoughts and impressions in a journal or online site. The more detail you use, the more you will appreciate it later on. Be sure to look around and notice details. This is easy at the start of the partial phase, when things may seem slow. But once you get to 90 percent of the sun being covered, the pace picks up.

This appeals to some more than others, but writing your thoughts during the eclipse can be a terrific way to record it. Be sure to complete the entry soon after the event so that you get details down before they fade from memory.

Camera with Natural Pinholes.

During a partial phase of the solar eclipse, an interesting setting for a photograph can be found under a tree that has its leaves projecting shadows. Refer back to chapter 11 for more detail.

Photography

If you are up for a real challenge, try photographing the eclipse. I call this a challenge for several reasons. First, the amount of light coming from the sun changes so much during the eclipse, and second, you don't have a chance to practice. You only get one shot at this. One more thing to think about: if you focus on taking pictures during the eclipse, after the eclipse, you may realize you saw it all through a viewfinder or screen and never saw the real thing. Please don't let that happen. Give yourself some time to soak in the corona without the camera between you and the sun.

If you want to take pictures, here are some recommendations. First, do not use the flash on your camera during any part of the eclipse. It does not help. Turn it off.

I do not recommend taking pictures of the sun with your phone. It will prove frustrating and disappointing. Use your phone camera only to take pictures of the surroundings, the horizon, and people looking up at the eclipse. This will get more difficult as the partial eclipse progresses because it will get darker and darker, and your camera will take longer and longer exposures to compensate for the lower light levels. This will make the images blurry, unless you are leaning on something stable or have a tripod.

ATTENTION: Never aim your camera at the sun during any of the partial eclipse. It may damage it beyond repair. And never look at the sun through a camera unless it has the proper filters. If you are the least bit unsure about f-stops, filters, and shutter speeds, you may want to leave photography to the pros. Take it all in, and enjoy the experience visually.

However, during totality you can aim your camera straight at the sun without fear of harm to it. For the total eclipse, I recommend a high-zoom camera. If you don't have a high-zoom camera, use what you've got. And, please, get yourself a tripod to mount the camera on. Spring for the twenty-five to fifty dollars for one. (My wife has had good luck locating this kind of thing at places like Goodwill Industries stores.) It doesn't guarantee great pictures, but without it, your images will suffer. If you have a digital single-lens reflex camera (if you don't recognize the term, you probably don't have one), you will need at least a 400mm lens to get a decent size image of the sun.

Before eclipse day, find the delayed shutter release (if it has one), and set it for two seconds or whatever the shortest interval is. Why? At high zoom, when you press the shutter release, a small vibration will occur, even on a tripod. The shutter delay lets the vibration subside before the shutter actually opens. NOTE: You usually have to set the shutter-release delay before each shot, which is a pain when you want to do it for every shot, but it is worth it. Practicing this maneuver (set shutter delay, press shutter release, wait for camera to take picture, repeat) will help things go smoothly during the eclipse.

If your camera has a manual setting and you want more detailed information on solar-eclipse photography, try these websites:

http://www.mreclipse.com/SEphoto/SEphoto.html
https://www.eclipse-chasers.com/photo/Photo.html
http://www.eclipse2017.org/2017/photographing.htm

This following site has some amazing images that required hundreds of separate images and hundreds of hours of post-processing to produce. Few people will be able to take images of this quality. But we can all gawk.

http://www.zam.fme.vutbr.cz/~druck/eclipse/

Eclipse Day

The most important thing is to get yourself to your viewing site on time. If you aren't in the path of the eclipse, you have missed it. Assuming you are where you want to be and the skies are clear, set up your gear. Make sure you know the current local time and the times of the four contacts of the eclipse. Then you are all set. Get ready for the big show.

1. Never look at the sun without proper eye protection.
2. Go see the eclipse, and take a kid with you.

Figure 20. A shoe box with a hole cut in one end and a sheet of white paper glued or taped on the inside of the opposite end. Wording on the box is optional.

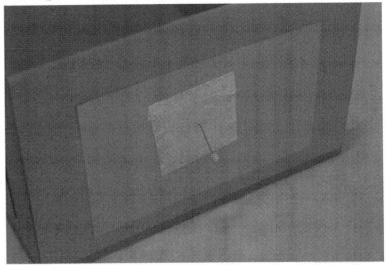

Figure 21. Aluminum foil is taped over the hole. Use a pin to make a small hole in the foil. Remove the pin. If the foil tears, use some of the extra you brought along. The sun's light will enter the box through this hole.

Figure 22. This is the pinhole seen from inside the box. The hole is round but not perfectly. That is okay.

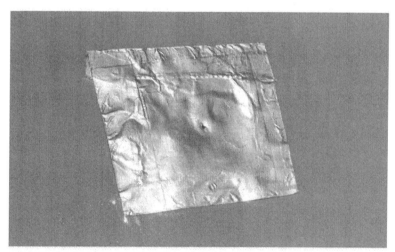

Figure 23. The foil and pinhole from the outside. This is the critical piece of the setup. Take good care of it.

Figure 24. To view the image of the sun, take the box outside and hold it with the foil end up, facing the sun. Don't you look up.

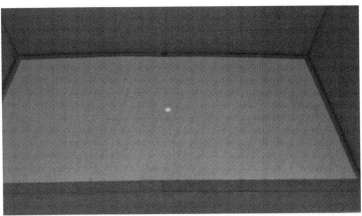

Figure 25. The image of the sun is the small, white circle. In this picture it is round. During the partial stage of the eclipse, it will gradually change to a crescent.

PART FIVE
A CALL TO ACTION

Making a Difference

I am passionate about this eclipse and most especially about children getting to see it. Consider this: if you, as an adult, decide to go see the eclipse, you can probably make it happen by juggling work schedules, joining with others to share expenses—whatever it takes to get the time and the means to experience this wonderful event. On the other hand, if a child wants to go see this eclipse, unless they live in the path of totality, they will need someone to take them. And so I ask the question, *whose responsibility is it to see that that happens?*

I believe that, ultimately, it is the responsibility of the parents of the child to see that that happens. They are the ones who seek the best for their child. They are the ones who have a vested interest in giving their child experiences that will enhance his or her life and allow their child to see and experience a wide variety of things. What I am asking here is for a greater commitment. Only some of you will be able to or be interested in doing this, but it may have a great and far-reaching effect.

If you live a long way from the eclipse, this will be a major undertaking for you. It will involve one or two overnights, travel, food, and lodging. One possible option is to include in your planning a child who otherwise would not be able to go see the eclipse. You may not live and move in circles where that is easy to discern. You may not know many or any families with school-age children who lack the means to travel to the eclipse. It may require some leg-work on your part to find someone who fits that description, but please try.

If you live closer to the path of totality, specifically in Oregon, Idaho, Wyoming, Nebraska, Kansas, Missouri, Illinois, Indiana, Kentucky, Tennessee, North Carolina, Georgia, or South Carolina, consider the fol-lowing action: meet with local leaders of youth clubs, schools, home educa-tors, or service organizations, and explore ways to take large numbers of students who live just outside the path into the path that day so they can see the TSE. It is a major fear of mine that children who live only ten to twenty miles outside the path of totality will not be given a ride or means to be inside the path that day and see this greatest of celestial sights.

Major and minor cities that have this issue to contend with include, but are not limited to, Eugene, Portland, and Bend, Oregon; Boise, Twin Falls, and Pocatello, Idaho; Laramie and Cheyenne, Wyoming; Columbus and Omaha, Nebraska; Manhattan, Topeka, and half of Kansas City, Kansas; half of Kansas City, Springfield, and half of St. Louis, Missouri; Springfield, Effingham, and Champaign, Illinois; Evansville, Indiana; Owensboro and Louisville, Kentucky; Franklin, Columbia, Chattanooga, and Knoxville, Tennessee; Atlanta, and all parts of northeast, Georgia; Asheville, Gastonia, and Charlotte, North Carolina; Spartanburg and Florence, South Carolina.

Consider this: in the path of totality, there are over 180 Kiwanis International Clubs. The mission of Kiwanis is to serve children. There are numerous other groups whose major goal is to work with children in some capacity. This is not an impossible mission. Tough and challenging, sure, but not impossible.

There are a couple more things to consider. The number of homeschooled children in the United States today is not huge but is growing and becom-ing more significant. Estimates vary, but two million is a figure cited by the National Home Education Research Institute in a January 3, 2011, report for 2010. I will use that figure for this discussion. Then based on my earlier estimate of sixty-five million people living within a day's trip of the centerline of the eclipse (20 percent of the nearly 320 million total population), we get about 400,000 homeschooled students who live within that distance. Of that, perhaps fifty to sixty thousand live in the path of totality. These are all esti-mates, not hard-and-fast, but they serve as working numbers.

What does this mean? What is the importance of this? Well, as I see it, the homeschool families that happen to live in the path of totality have a rare opportunity to help like-minded people view the eclipse in a comfortable, safe environment. The families living in the path likely know the area where they live and can arrange with local churches and other organizations to use fields, parking lots, rest rooms, and even food concessions to make this an event to remember. There are many homeschool organizations in the affected states, perhaps providing a means of communication for the families involved in developing these activities and locations.

And this is just scratching the surface of what could be done. As 2017 approaches and more people become aware of the eclipse, its educational potential and the opportunity to network and share with others should increase. People will no doubt come up with their own creative alternatives and solutions that far exceed what I could come up with by myself. In this case, I want to be the instigator.

Not just homeschoolers live in the eclipse path. Lots of others with no particular involvement or association with a school live there. Senior citizens, empty nesters, singles, companies, organizations—you name it. They each can play a role in providing resources for the many visitors sure to come that day. As August 21, 2017, approaches, interest in the eclipse will rise. I predict, and this takes no genius on my part, that this is going to be big. This eclipse may, in fact, be seen by more people than any eclipse in history outside of China and India. (China and India have high concentrations of large populations.) Millions of people will travel—some, great distances—to experience it. The hospitality industry needs to get ready, but so do locals. I can see churches, schools, fraternal organizations, and parks in the path of totality setting up fields for parking and selling refreshments and allowing use of restrooms for this event. Some people will make day trips, but many will choose to stay a night or two in the area.

The greatest influx will likely happen in regions where there is a populated area just outside the eclipse path and an interstate highway directly connecting it to a town in the eclipse path. Examples are the Portland, Oregon, area toward Salem on Interstate 5; the Salt Lake City area toward Idaho Falls on I-15; the Denver and Cheyenne areas toward Casper and

Douglas, Wyoming, via I-25; the Omaha area toward Lincoln via I-80; the Knoxville and Chattanooga, Tennessee, areas toward Sweetwater and Athens on I-75; the Atlanta, Georgia, area toward many places along I-85; the Charlotte, North Carolina, to Greenville, South Carolina, area close to I-85 or Columbia, South Carolina, via I-77; and the Savannah, Georgia, and Jacksonville, Florida, areas up I-95 to anyplace in South Carolina that's in the path.

How Can You Become Involved?

Now that's a hard question. What can you do to help someone experience the eclipse? One goal of this book is to get as many school-age children to this eclipse as possible. Are you a member of a civic organization that can help? Do you live in a community that has such organizations? Are they aware of the eclipse and what they can do? Could you encourage them to get involved?

Do you have children between the ages six to sixteen? Do you have nieces, nephews, cousins, grandchildren, godchildren, or family members who do? Get to work. Plan a trip. Spread the word. Encourage everyone in your sphere of influence to consider going to experience this TSE. You would be surprised what one person can do. Please consider doing something to give yourself and a child this once-in-a-lifetime experience. You will not regret it.

As I have said, this eclipse is going to happen, and it is going to happen on August 21, 2017. It is a rare event in the United States. Everyone should see it, but especially kids. We are being presented with a great opportunity. We should not squander it. Will it be inconvenient? For many, yes. Will it cost time and money? Some, yes. But will the time and money spent be worth it? You will have to answer that for yourself after the fact, but I have no doubt that the answers are yes and yes. Let's work to make this happen.

1. Never look at the sun without proper eye protection.
2. Go see the eclipse, and take a kid with you.

PART SIX
WHAT'S NEXT?

I recommend that children younger than about four years old not view this eclipse. With children younger than that, viewing the eclipse should be decided on a case-by-case basis, with safety being the most important consideration. The child's temperament will play a big factor in the decision. Is the child compliant? Can you trust the child to do what you say and not look at the sun? That is a parent's job, and parents know their child better than anyone.

That being said, have I left those parents of very young children out in the cold? Have they no recourse for viewing this rare event? Is this a once-in-a-lifetime event that will be lost to them? And are there no good options available to people who are just now expecting their first child? Will they have to wait another forty years until the next TSE graces our land?

The answer to that is an unqualified *no*. The fact is that another total solar eclipse will grace the United States within a much shorter time span than we might expect. In a bit of unfairness that the solar system seems to exhibit (not that it really cares, of course), two total solar eclipses cover America within seven years. So mark your calendars for Monday, April 8, 2024.

That one will have a different path that will carry it north from Mexico through Texas, Oklahoma, Arkansas, Missouri, Illinois, Kentucky,

Indiana, Ohio, Pennsylvania, New York, Vermont, New Hampshire, Maine, and into Canada. Major cities in the path include Austin, Dallas, and Fort Worth, Texas; Little Rock, Arkansas; Evansville, Bloomington, and Indianapolis, Indiana; Toledo, Dayton, Springfield, and Cleveland, Ohio; Erie, Pennsylvania; Buffalo, Rochester, and Syracuse, New York; Burlington, Vermont; Hamilton and Saint Catherine, Ontario; Montreal and Sherbrooke, Quebec; and Fredericton, New Brunswick.

Within two hundred miles of the centerline are San Antonio, Oklahoma City, St. Louis, Memphis, Nashville, Louisville, Chicago, Detroit, Cincinnati, Columbus, Pittsburgh, Albany, and Boston in the United States and Toronto and Ottawa in Canada. This eclipse will be longer (with over four minutes of totality at its greatest) and the path of totality wider (over 110 miles).

This means that children age four and under for the 2017 eclipse will be from seven to twelve years old for the one in 2024—prime ages for eclipse viewing. So set your sights for then. The time will pass before you know it.

If you read this book before the eclipse of 2017 and have older children who cannot wait until then to see one, there is an option for you.

Date	Location
March 9, 2016	Indonesia and Pacific Ocean

After the eclipse of August 21, 2017, the next several eclipses are listed here.

Date	Location
July 2, 2019	Pacific Ocean, Chile, and Argentina
December 14, 2020	Pacific to the Atlantic, again across Chile and Argentina
December 4, 2021	Antarctica
April 20, 2023	Indian and Pacific Oceans, East Timor, and New Guinea

See what I mean about having to travel to see eclipses? So if you enjoy traveling and viewing eclipses, you have just been given a planning guide for excursions for the next several years. Your journey starts here.

APPENDICES

Appendix I
Glossary

Here is a list of some words and their meanings that are useful to know when talking about eclipses. Although this book is aimed at nonastronomers, every discipline has its own words with specialized meanings. This list forms a starting point for becoming conversant in eclipse-ese.

Altitude. Height of an object above the horizon, measured in degrees. Something straight up has an altitude of 90 degrees. Something on the horizon has 0 degrees altitude.

Azimuth. The direction you face on the horizon. An example is a ship's heading. It is measured in degrees from due north. You can use a compass to measure azimuth. North has an azimuth of 0 degrees; east, 90 degrees; south, 180 degrees; and west, 270 degrees.

Centerline. The center of the line of the path of totality. Maximum duration of the eclipse occurs on this line.

Contact. When the moon appears to touch an edge of the sun. First contact is the initial bite the moon appears to take out of the sun. Second contact is the appearance of the first diamond ring and the beginning of totality. Third contact is the appearance of the second diamond ring and the end of totality. Fourth contact is when the last bit of the moon is no longer seen as it leaves the sun's bright surface.

Eclipse. An occultation involving the sun, moon, and earth.

Magnitude of a solar eclipse. The depth of the eclipse. Magnitude is described as the fraction of the solar diameter that is covered by the moon.

A magnitude of one or greater signifies a total solar eclipse. A magnitude less than one but greater than zero is a partial solar eclipse.

Occultation. When one object passes in front of another. The nearer object is said to occult, or hide, the distant one.

Path of the eclipse (a.k.a., the path of totality). The shadow (or umbra) of the moon touches the earth in the shape of an oval that is typically fifty to one hundred miles across. As the moon orbits the earth, its shadow moves across the earth's surface, making a line fifty to one hundred miles wide. Every point in that broad line will experience a total solar eclipse.

Penumbra. The outer part of a shadow when the light source is only partially obscured. This results in a partial eclipse.

Photosphere. The apparent surface of the sun. The sun is gas and plasma. No part of it is solid. But the light generated inside it has to work its way to a layer that changes from opaque to transparent. The sun's light appears to come from there. We call it the sphere of light or photosphere.

Solar eclipse. When the moon passes between the sun and the earth.

Solar shades. Also known as eclipse-viewing glasses, solar shades are usually inexpensively made but effective. They are safety devices worn like glasses for viewing the partially eclipsed sun (or the sun on any non-eclipse day, too). Be sure to use them properly when viewing the sun.

Totality. The brief period of time when the sun is completely eclipsed.

TSE. I use this throughout the book. It is the abbreviation for total solar eclipse.

Umbra. The darker part of a shadow where the light source is completely obscured.

Universal Time (UT). Standard time at Greenwich, England. Formerly based on the rotation of the earth. Now based on the atomic clock time known as UTC.

Remember:
1. Never look at the sun without proper eye protection.
2. Go see the eclipse, and take a kid with you.

Appendix II
Eclipse Timetables

Here are timetables of the total solar eclipse of August 21, 2017, giving the circumstances of the eclipse as the path enters each state from the west. These use a twenty-four-hour clock. See chapter 10 for details on understanding these.

Oregon

Event		Time (PDT)	Alt	Azi
Start of partial eclipse	(C1) :	9:04:30.7	26.9°	100.3°
Start of total eclipse	(C2) :	10:15:55.2	38.9°	115.5°
End of total eclipse	(C3) :	10:17:53.7	39.3°	116.0°
End of partial eclipse	(C4) :	11:36:03.9	50.3°	138.2°

Idaho

Event		Time (MDT)	Alt	Azi
Start of partial eclipse	(C1) :	10:10:07.2	32.7°	106.7°
Start of total eclipse	(C2) :	11:24:54.6	44.8°	124.5°
End of total eclipse	(C3) :	11:27:04.2	45.1°	125.1°
End of partial eclipse	(C4) :	12:48:01.6	54.7°	152.3°

Wyoming

Event		Time (MDT)	Alt	Azi
Start of partial eclipse	(C1) :	10:16:27.2	38.2°	112.9°
Start of total eclipse	(C2) :	11:34:23.8	50.0°	134.0°
End of total eclipse	(C3) :	11:36:43.1	50.3°	134.8°
End of partial eclipse	(C4) :	12:59:48.2	57.6°	167.4°

Nebraska

Event		Time (MDT)	Alt	Azi
Start of partial eclipse	(C1) :	10:25:21.1	44.8°	120.9°
Start of total eclipse	(C2) :	11:46:52.0	55.7°	147.4°
End of total eclipse	(C3) :	11:49:21.1	55.9°	148.4°
End of partial eclipse	(C4) :	13:14:09.9	59.3°	187.2°

Kansas

Event	Time (CDT)	Alt	Azi
Start of partial eclipse (C1) :	11:39:14.1	53.4°	132.8°
Start of total eclipse (C2) :	13:04:42.4	61.5°	169.4°
End of total eclipse (C3) :	13:07:20.1	61.6°	170.7°
End of partial eclipse (C4) :	14:32:51.5	58.1°	212.6°

Missouri

Event	Time (CDT)	Alt	Azi
Start of partial eclipse (C1) :	11:40:36.6	54.2°	134.0°
Start of total eclipse (C2) :	13:06:22.8	61.9°	171.6°
End of total eclipse (C3) :	13:09:01.0	61.9°	173.0°
End of partial eclipse (C4) :	14:34:30.8	57.9°	214.7°

Illinois

Event	Time (CDT)	Alt	Azi
Start of partial eclipse (C1) :	11:51:06.6	59.5°	143.9°
Start of total eclipse (C2) :	13:18:39.6	63.7°	189.1°
End of total eclipse (C3) :	13:21:19.6	63.6°	190.6°
End of partial eclipse (C4) :	14:46:13.0	55.1°	228.6°

Kentucky

Event	Time (CDT)	Alt	Azi
Start of partial eclipse (C1) :	11:54:05.5	60.9°	146.9°
Start of total eclipse (C2) :	13:21:59.7	63.9°	194.0°
End of total eclipse (C3) :	13:24:39.7	63.8°	195.5°
End of partial eclipse (C4) :	14:49:16.1	54.2°	232.0°

Tennessee

Event	Time (CDT)	Alt	Azi
Start of partial eclipse (C1) :	11:57:51.7	62.5°	151.0°
Start of total eclipse (C2) :	13:26:07.4	64.0°	200.1°
End of total eclipse (C3) :	13:28:47.1	63.8°	201.5°
End of partial eclipse (C4) :	14:52:58.5	53.0°	235.8°

North Carolina

Event	Time (EDT)	Alt	Azi
Start of partial eclipse (C1) :	13:05:06.7	65.3°	159.7°
Start of total eclipse (C2) :	14:33:47.5	63.5°	211.2°
End of total eclipse (C3) :	14:36:25.9	63.2°	212.5°
End of partial eclipse (C4) :	15:59:40.2	50.5°	242.3°

Georgia

Event	Time (EDT)	Alt	Azi
Start of partial eclipse (C1) :	13:06:55.5	65.9°	162.1°
Start of total eclipse (C2) :	14:35:39.4	63.2°	213.8°
End of total eclipse (C3) :	14:38:17.3	62.9°	215.0°
End of partial eclipse (C4) :	16:01:15.8	49.9°	243.8°

South Carolina

Event	Time (EDT)	Alt	Azi
Start of partial eclipse (C1) :	13:07:23.1	66.1	162.7°
Start of total eclipse (C2) :	14:36:07.6	63.2	214.4°
End of total eclipse (C3) :	14:38:45.4	62.9	215.7°
End of partial eclipse (C4) :	16:01:39.7	49.7	244.1°

Appendix III
Communities in the Path of Totality

To help you understand the scope of this eclipse, here is a list of the towns and cities with over three thousand people that are in the path of totality. The list is by state, in the order the moon's shadow touches each one, from west to east. Town names are listed alphabetically. Smaller towns are included if they are close to the centerline of the eclipse.

There are five state capitals in the path of totality: Salem, Oregon; Lincoln, Nebraska; Jefferson City, Missouri; Nashville, Tennessee; and Columbia, South Carolina.

Oregon

Albany	Lebanon	Prineville
Baker City	Lincoln City	Redmond
Corvallis	Madras	Salem
Dallas	McMinnville	Sheridan
Four Corners	Molalla	Silverton
Hayesville	Monmouth	Stayton
Independence	Mt. Angel	Sweet Home
Jefferson	Newport	Toledo
John Day	Ontario	Warm Springs
Keizer	Philomath	Woodburn

Idaho

Ammon	Idaho Falls	Rigby
Driggs	Ketchum	Shelley
Emmett	Payette	St. Anthony
Fruitland	Rexburg	Weiser

Wyoming

Casper	Jackson	Thermopolis
Douglas	Lander	Torrington
Dobois	Riverton	Wheatland
Glenrock		

Nebraska

Alliance	Geneva	North Platte
Auburn	Gering	Ravenna
Aurora	Gothenburg	Sabetha
Beatrice	Grand Island	Scottsbluff
Broken Bow	Hastings	Seward
Central City	Kearney	Sutton
Cozad	Lexington	Tecumseh
Fairbury	Lincoln	Wilber
Falls City	Minden	York

Nebraska

Atchison	Kansas City (in Part)	Leavenworth
Hiawatha	Lansing	

Missouri

Ashland	Higginsville	Parkville
Ballwin	Holts Summit	Perryville
Boonville	Independence	Richmond
California	Jackson	Riverside
Cameron	Jefferson City	Savannah
Cape Girardeau	Kansas City (in part)	Scott City
Carrollton	Kearney	Sedalia
Centralia	Lexington	St. Clair
Chesterfield	Liberty	St. Genevieve
Chillicothe	Manchester	St. Joseph
Columbia	Maplewood	St. Louis (in part)
Crystal City	Marshall	Sullivan
De Soto	Mehlville	Tipton
Eureka	Mexico	Union
Farmington	Moberly	Valley Park
Fenton	North Kansas City	Warrenton
Festus	Odessa	Washington
Fredericktown	Pacific	Wentzville
Fulton	Park Hills	

Illinois

Anna	DuQuoin	Murphysboro
Belleville	Dupo	Pinckneyville
Benton	Eldorado	Red Bud
Carbondale	Freeburg	Sparta
Carterville	Harrisburg	Waterloo
Chester	Johnston City	West Frankfort
Columbia	Marion	

Kentucky

Benton	Greenville	Paducah
Bowling Green	Hopkinsville	Princeton
Central City	Madisonville	Russellville
Dawson Springs	Marion	Scottsville
Franklin		

Tennessee

Alcoa	Harriman	Mt. Juliet
Ashland City	Henderson	Murfreesboro
Athens	Kingston	Nashville
Clarksville	La Vergne	Oak Ridge
Cleveland	Lafayette	Portland
Cookeville	Lebanon	Smithville
Crossville	Lenoir City	Smyrna
Dayton	Livingston	Sparta
Etowah	Loudon	Springfield
Farragut	Madisonville	Sweetwater
Gallatin	Maryville	White House
Goodlettsville	McMinnville	

Georgia

Clayton	Hartwell	Toccoa
Cornelia		

North Carolina

Andrews	Cullowhee	Murphy
Brevard	Franklin	

South Carolina

Abbeville	Georgetown	North Charleston
Anderson	Goose Creek	Orangeburg
Andrews	Greenville	Pendleton
Bamberg	Greenwood	Pickens
Batesburg-Leesville	Greer	Piedmont
Belton	Hanahan	Saluda
Cayce	Honea Path	Seneca
Central	Irmo	Simpsonville
Charleston	Isle of Palms	Summerville
Clemson	Kingstree	Sumter
Clinton	Ladson	Taylors
Columbia	Laurens	Travelers Rest
Denmark	Lexington	Union
Dentsville	Liberty	Walhalla
Easley	Lugoff	Williamston
Edgefield	Manning	Winnsboro
Fountain Inn	Moncks Corner	Woodruff
Gantt	Newberry	

Here is a list of larger cities NOT IN but NEAR the path of totality (no more than 200 miles away, except for Seattle). This distance would require travel but could be a day trip or one overnight. Some of the cities are from states outside the path of totality.

Oregon	South Dakota	Kentucky
Eugene	Rapid City	Lexington
Portland	Sioux Falls	Louisville
Washington	**Nebraska**	**Ohio**
Seattle (300 miles)	Norfolk	Cincinnati
Tacoma	Omaha	
		Tennessee
Idaho	**Kansas**	Chattanooga
Boise	Manhattan	Jackson
Pocatello	Topeka	Knoxville
	Wichita	Memphis
Montana		
Billings	**Missouri**	**Alabama**
Butte	Springfield	Huntsville
Utah	**Iowa**	**Georgia**
Logan	Cedar Rapids	Atlanta
Salt Lake City	Des Moines	Augusta
	Sioux City	Macon
Wyoming		Savannah
Cheyenne	**Illinois**	
	Springfield	**North Carolina**
Colorado		Asheville
Boulder	**Indiana**	Charlotte
Denver	Evansville	Greensboro
Fort Collins	Indianapolis	Raleigh
		Florida
		Jacksonville

Appendix IV
State Parks in the Path of Totality

The states are listed chronologically as the path of totality touches them. Within the states, the parks are alphabetical.

OREGON
Agate Beach State Park
Beverly Beach State Park
Boiler Bay State Wayside
Brian Booth State Park, a.k.a., Ona Beach State Park
Cape Kiwandi State Natural Area
Cascadia State Park
Clyde Holliday State Recreation Area
D River State Recreation Area
Depot Bay State Park
Devil's Lake State Park
Devil's Punch Bowl State Park
Driftwood Beach State Recreation Area
Fogarty Creek State Recreation Area
Gleneden Beach State Wayside
Lost Creek State Park
Maud Williamson State Recreation Area
Peter Skene Ogden State Park
Road's End State Park

Sarah Helmick State Park
Seal Rock State Wayside
Silver Falls State Park
Smith Rock State Park
South Beach State Park
The Cove Palisades State Park
Unity Lake State Recreation Area
Willamette Mission State Park
Yaquina Bay State Recreation Area

IDAHO
Lake Cascade State Park
Land of the Yankee Fork State Park

WYOMING
Boysen State Park
Edness Kimball Wilkins State Park
Glendo State Park
Guernsey State Park

NEBRASKA
Arnold State Recreation Area
Big Lake State Park
Bluestem Lake State Recreation Area
Box Butte Reservoir State Recreation Area
Conestoga Lake State Recreation Area
Indian Cave State Park
Kirkman's Cove State Recreation Area
Lake Minitare State Recreation Area
Pawnee State Recreation Area
Rockford Lake State Recreation Area
Stagecoach Lake State Recreation Area
Wagon Train Lake State Recreation Area

KANSAS
Atchison County State Park
Brown County State Park
Nemaha County State Park

MISSOURI
Big Lake State Park
Binder State Park
Finger Lakes State Park
Graham Cave State Park
Hawn State Park
Lewis and Clark State Park
Onondaga Cave State Park
Robertsville State Park
Trail of Tears State Park
Van Meter State Park
Wallace State Park

ILLINOIS
Cave-in Rock State Park
Dixon Springs State Park
Ferne Clyffe State Park
Fort Massac State Park
Giant City State Park
Lake Murphysboro State Park

KENTUCKY
Kentucky Dam Village State Resort Park
Lake Malone State Park
Mineral Mound State Park
Pennyrile Forest State Resort Park

TENNESSEE
Bledsoe Creek Camping State Park
Cedars of Lebanon State Park
Cumberland Mountain State Park
Dunbar Cave State Park
Edgar Evins State Park
Fall Creek Falls State Park
Hiawasee/Ocoee State Park
Long Hunter State Park
Port Royal State Park
Rock Island State Park
Sellars Farm State Park

NORTH CAROLINA
Gorges State Park

GEORGIA
Black Rock Mountain State Park
Moccasin Creek State Park
Tallulah Gorge State Park
Tugaloo State Park
Unicoi State Park

SOUTH CAROLINA
Aiken State Park
Baker Creek State Park
Caesar's Head State Park
Charles Towne Landing State Historic Site
Colleton State Park
Colonial Dorchester State Historic Site
Devils Fork State Park
Dreher Island State Park
Givhans Ferry State Park

Hampton Plantation State Historic Site
Jones Gap State Park
Keowee-Toxaway State Park
Lake Greenwood State Park
Lake Hartwell State Park
Musgrove Mill State Historic Site
Oconee State Park
Oconee Station State Historic Site
Paris Mountain State Park
Poinsett State Park
Rose Hill Plantation State Historic Site
Sadlers Creek State Park
Santee State Park
Sesquicentennial State Park
Table Rock State Park

Appendix V

Interstate Highway Rest Areas in the Path of Totality

Something to Consider.

One possibility for a viewing site is an interstate highway rest area. They are easy to get to and usually have a fair amount of parking, as well as restrooms. Some even have vending machines with beverages and snacks. The following table has a list of interstate highway rest areas within the path of totality. Some, however, do not allow overnight parking. Be sure you abide by the rules for any given rest area. The list is organized by locations, starting in the west and going east.

Table Headings:

Col. 1 – Interstate highway number

Col. 2 – State name

Col. 3 – Highway mile marker

Col. 4 – Comments (These are all rest areas in both directions, unless otherwise noted.)

Col. 5 – The duration of totality in minutes and seconds

Col. 1	Col. 2	Col. 3	Col. 4	Col. 5
5	OR	240		1:59
84	OR	335		2:07
84	ID	142	turn out only, no facilities	2:17
25	WY	171	turn out only, no facilities	2:24
25	WY	126		2:26
25	WY	92		1:56
80	NE	269	eastbound	1:37
80	NE	270	westbound	1:41
80	NE	314	eastbound	2:35
80	NE	316	westbound	2:35
80	NE	350	eastbound	2:27
80	NE	355	westbound	2:24
80	NE	375	eastbound	2:05
80	NE	381	westbound	1:57
29	MO	108	southbound	1:43

29	MO	82		2:22
29	MO	27		2:22
35	MO	34	northbound	2:37
35	MO	35	southbound	2:37
70	MO	57		2:13
70	MO	104		2:38
70	MO	167	eastbound	2:10
70	MO	169	westbound	2:06
70	MO	198		1:32
44	MO	235		2:39
55	MO	110		2:14
55	MO	160		2:40
57	IL	32		2:37
24	IL	37	welcome center	2:26
24	KY	7	welcome center	2:11
24	KY	93	welcome center	2:32

Appendix VI
Reproducible Puzzles and Games

Double Puzzle

Use the scrambled letters below on the left to make an eclipse related word and write in the squares on the right. Then take the letters that appear in ⬭ boxes and unscramble them, using the final clue, for the final message.

FINAL CLUE: It turns day into night.

CONRAO

RIBHYD

LUNRAAN

PATRILA

RAUPNMEB

NRCETSEC

Answers can be found after the next puzzle.

Word Find

ECLIPSE

How many words of three letters or longer can you make from the word ECLIPSE?

We found forty-five. If you find 1–10, go back to school; 11–20, average; 21–30, looking good; more than thirty, you are a wordsmith.

_____

_____

_____

Answers can be found on the next page.

Puzzle Answers

Double Puzzle:
ANSWERS: CORONA, HYBRID, ANNULAR, PARTIAL, PENUMBRA, CRESCENT
FINAL ANSWER: TOTAL ECLIPSE

Word Find:

eli	eel	lei	lee	lie	lip
ice	pee	pie	pis	see	sic
sip	else	epic	eels	clip	leis
lees	lies	lice	lips	lisp	ices
isle	peel	pees	pies	pile	seep
slip	spec	epics	clips	peels	plies
piece	piles	sleep	slice	spiel	spice
pieces	specie	splice			

The words three letters or longer are:

Word-Search Puzzles

In our word searches, words are placed horizontally, vertically, and diagonally, both forward and from back to front. There are also lots of overlaps between words, so you will need a keen eye to spot all the words and solve the puzzles.

All these puzzles have the names of towns and cities in the path of totality for the total solar eclipse of August 21, 2017.

Oregon

```
Y I S Z Y O N T A R I O Q O S
S N Q I N O T Y A T S M H N H
M W M Y L K E I Z E R A P O E
L I N C O L N C I T Y B H S R
W P M R M I A T L E W T A R I
A N V O C I R V S U A S L E D
R G O T N O N V R M J A L F A
M Y F N P M I N O O R L A F N
S E A W A L O L V N C L L E R
P H E X L B I U Y I J A O J E
R N I E Y H E C T N L D M N D
I U V W P B N L U H A L B C M
N S I L V E R T O N R B E P O
G N R U B D O O W A H E L F N
S E L L I V E N I R P B U A D
```

ALBANY	CORVALLIS
DALLAS	HAYESVILLE
JEFFERSON	KEIZER
LEBANON	LINCOLNCITY
MCMINNVILLE	MOLALLA
MONMOUTH	NEWPORT
ONTARIO	PHILOMATH
PRINEVILLE	REDMOND
SHERIDAN	SILVERTON
STAYTON	WARMSPRINGS
WOODBURN	

The solution is on the next page.

Oregon Solution

```
+  +  S  +  +  O  N  T  A  R  I  O  +  +  S
+  +  +  I  N  O  T  Y  A  T  S  +  H  N  H
+  +  M  +  L  K  E  I  Z  E  R  A  +  O  E
L  I  N  C  O  L  N  C  I  T  Y  +  H  S  R
W  +  M  +  M  +  A  T  +  E  +  T  A  R  I
A  N  +  O  +  I  R  V  S  +  A  S  L  E  D
R  +  O  +  N  O  N  V  R  M  +  A  L  F  A
M  +  +  N  P  M  I  N  O  O  +  L  A  F  N
S  +  +  W  A  L  O  L  V  +  C  L  L  E  R
P  +  E  +  L  B  I  U  Y  I  +  A  O  J  E
R  N  +  E  +  H  E  +  T  N  L  D  M  +  D
I  +  +  +  P  +  +  L  +  H  A  L  +  +  M
N  S  I  L  V  E  R  T  O  N  +  B  E  +  O
G  N  R  U  B  D  O  O  W  +  +  +  L  +  N
S  E  L  L  I  V  E  N  I  R  P  +  +  A  D
```

(Over, Down, Direction)

ALBANY (14, 15, NW)   CORVALLIS (11, 9, NW)
DALLAS (12, 11, N)   HAYESVILLE (13, 2, SW)
JEFFERSON (14, 10, N)   KEIZER (6, 3, E)
LEBANON (8, 12, NW)   LINCOLNCITY (1, 4, E)
MCMINNVILLE (3, 3, SE)   MOLALLA (13, 11, N)
MONMOUTH (3, 5, SE)   NEWPORT (2, 11, NE)
ONTARIO (6, 1, E)   PHILOMATH (5, 12, NE)
PRINEVILLE (11, 15, W)   REDMOND (15, 9, S)
SHERIDAN (15, 1, S)   SILVERTON (2, 13, E)
STAYTON (11, 2, W)   WARMSPRINGS (1, 5, S)
WOODBURN (9, 14, W)

More reproducible puzzles are available at this book's companion website: http://goseetheeclipse.com.

Citations

Eclipse maps courtesy of Fred Espenak, NASA's Goddard Space Flight Center.

The maps used in this book are provided by Google Maps. See individual maps for attribution. The eclipse paths are provided by Xavier Jubier, using data from NASA, and Fred Espanek, Goddard Space Flight Center.

For information and predictions for solar and lunar eclipses, visit Fred Espenak's website, www.EclipseWise.com, and Xavier Jubier's website, http://xjubier.free.fr/en/index_en.html.

About the Author

Chap Percival lives in Sarasota, Florida with his wife, Bonnie. He has a daughter, a son-in-law and a granddaughter. He was born in Scranton, Pennsylvania. He has a BA in mathematics and physics from Taylor University, an MAT in planetarium education from Michigan State University, and an MEd in instructional technology from the University of Virginia.

Chap has been involved in astronomy education since 1969 as a classroom teacher, club sponsor, and planetarium director. He has published articles for local newspapers and has been interviewed on radio. He has volunteered with the National Parks Service as a sky interpreter. He also gave an astronomy talk at the Grand Canyon Star Party and has led groups to view five different total solar eclipses.

Chap is a lifelong educator and began teaching astronomy at Pine View School in Osprey, Florida, one of the top high schools in the nation, in 1995.

He has a passion for sharing his knowledge of things in the sky that amaze, astound, and awe him.

The author is available for interviews, speaking engagements, and consulting. Contact him at goseetheeclipse@gmail.com.

Testimonials

Eloise said: "Seeing a total solar eclipse will always be one of the highlights of all my many travel experiences. Having Chap Percival as my astronomy expert made it an even more valuable, educational, and fun experience."

Nancy said: "Viewing a total solar eclipse with Chap was a great adventure. He loves to share his wealth of knowledge, but at the same time, he makes it fun. This is an event you don't want to miss."

Jane said: "We have been on two trips to view a total solar eclipse with Chap. He is extremely knowledgeable about the event, so the trips were educational. He also included many interesting extras, so we had a lot of fun. Great time! We would highly recommend it."

Joyce said: "In 1999, I had the opportunity to travel to Hungary with a group to view the total solar eclipse. The tour was arranged and led by Chap Percival, a science teacher from Florida. Although I am a now retired social-studies teacher from Minnesota, the experience was something I loved sharing with my students upon my return. They were fascinated, along with the science teachers in my high school, who enjoyed using the video that Chap produced in their classes. The trip was well organized and included many other interesting sights and experiences."

Made in the USA
San Bernardino, CA
28 April 2017